OPERATING SYSTEMS

OPERATING SYSTEMS

An Introduction

R. GARG
G. VERMA

MERCURY LEARNING AND INFORMATION
Dulles, Virginia
Boston, Massachusetts
New Delhi

Publisher: David Pallai
MERCURY LEARNING AND INFORMATION
22841 Quicksilver Drive
Dulles, VA 20166
info@merclearning.com
www.merclearning.com
1-800-232-0223

R. Garg and G. Verma. *Operating Systems: An Introduction.*
ISBN: 978-1-942270-38-6

Library of Congress Control Number: 2015957712

171819321   Printed in the USA on acid-free paper.

CONTENTS

CHAPTER 1

OPERATING SYSTEMS: OVERVIEW

Purpose of this Chapter

Computer hardware cannot perform any task without the presence of software. Software gives instructions to the computer hardware to perform desired actions. The operating system is software that provides an environment for the user to perform tasks on the computer system. The purpose of this chapter is to provide a basic understanding of the operating system, its goals and classification.

INTRODUCTION

If there are two people who speak different languages, then how will they communicate with each other?

If person A speaks German and person B speaks English, then it will be very difficult for them to converse with each other. The solution will be to have another person in between, who understands and can communicate in both languages. This person in between is called an *intermediate* who will translate the communication for persons A and B.

The same scenario also occurs in computer systems. On one side there is the hardware of the computer system, and a user on the other side. Without an intermediate between them it is not possible for the user to operate the computer system efficiently. This intermediate in between the computer hardware and user is called *operating system* (OS). Thus, the operating system can be defined as an intermediate between computer hardware and user that provides an environment for the user so that he can use the hardware in an efficient and convenient manner.

OS can also be defined as an *interface* between the user and the computer hardware.

CONCEPT OF OPERATING SYSTEM

A computer system is considered to be a combination of many compo- nents, like hardware, the operating system, the application programs and the users. *Hardware* means the physical parts of the computer system. It is comprised of input and output devices, the *central processing unit* (CPU), memory devices, etc.

An *operating system* is a collection of software that manages computer hardware resources and provides common services to computer programs. The operating system is one of the main components in a computer system. In technical terms, it is software that manages the physical hardware of the system. An operating system controls the allocation of resources and services, such as memory, processors, devices, and information.

Application programs are programs that enable the end user to perform some specific, productive tasks, for example word processing, spreadsheets, Paint, etc. Application software is software that interacts with end users and performs what- ever the end user is working on, for example spreadsheet software, a photo appli- cation, etc. Application software does not deal with the architecture of a computer.

System software is software that manages and controls the computer hardware so that the application software can perform a task. System software is different from application software. System software performs tasks like transferring data from the computer memory to disc or displaying text on an output device. Specific kinds of system software include loading programs, operating systems, device drivers, programming tools, compilers, assemblers, linkers, and utility software. System soft- ware is software that deals directly with the architecture of the computer hardware.

End users are the human beings who use the computer system for performing their tasks.

The various components of a computer system are shown in Figure 1.1.

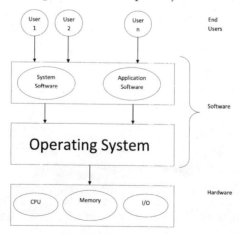

FIGURE 1.1. Main components of a computer system

GOALS OF AN OPERATING SYSTEM

An operating system is an important part of every computer system. It performs various tasks and thus it has various goals. It is a program that acts as an interface between the user and the computer hardware and controls the execution of all kinds of programs. The main reason for the existence of an operating system is to make things easier for the user of a computer system.

An OS has three main goals, which are as follows:

1. Optimum and efficient use of the computer system hardware
2. To provide the user with convenience and user friendliness
3. Hide the complexity of the computer hardware

According to the type of user, the goal of the OS is set. For example,

- A personal computer in a home to be used by family members is mainly for playing games, watching videos, and Internet surfing. In such systems the OS which is in use should be user friendly. The user should be able to use the system easily without any complexity or without any support of tutorial or training.
- On the other hand, if the system is to be used in organizations where its resources are to be shared among various users, then the OS should be capable enough to provide the effective and optimum utilization of the computer system and its resources.
- The OS hides the complexity of computer hardware from the user and provides only an abstract view of it. Complexity of hardware is hidden from the user so that it is convenient and easy for the user to use the system.

In short, the main goal of the OS depends mainly upon the type of user and purpose of the computer system. Thus, the OS can also be defined as a set of system programs that provide the above mentioned goals to the user.

Every computer system has various resources as shown in Figure 1.2.

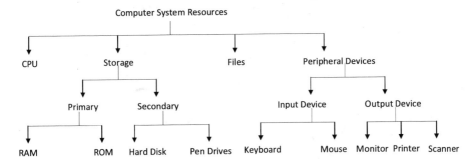

FIGURE 1.2. Various resources of the computer system

These resources are to be managed and their controlled access is to be provided to the processes, devices, and users. Thus, the OS is a resource manager that manages all the resources like memory, input/output devices and the CPU of the system, and provides their controlled access to the user. Users and processes access the computer's resources through the operating system which is shown in Figure 1.3.

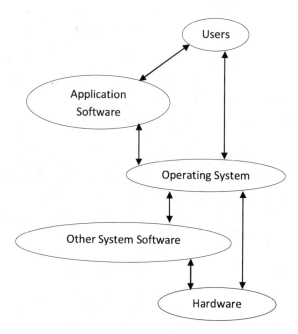

FIGURE 1.3. Presence of an operating system between user and hardware

As mentioned, an operating system is called an intermediary agent between the user and the computer hardware, that:

■ Manages the computer's resources (hardware, abstract resources, software)
■ Is a resource allocator
■ Is also used to control programs to prevent errors and improper computer use
■ Is "interrupt" driven. The signals that are sent by devices to the CPU are called *interrupts*

These interrupts are usually input/output devices. They signal the CPU to halt its current activities and execute the appropriate part of the operating system. There are three types of interrupts:

■ Hardware interrupts
■ Software interrupts
■ Traps

Let's discuss each one of them:

1. **Hardware Interrupt:** The interrupts that are generated by hardware devices to indicate that they need some attention from the OS. For example, they have just completed a task that the operating system previously requested, such as transferring data between the hard drive and memory;
2. **Software Interrupts:** The interrupts that are generated by programs when they want to request a *system call* to be performed by the operating system;
3. **Traps:** *Traps* are the signals that are generated by the CPU itself to indicate that some error or condition occurred for which assistance from the operating system is needed.

Interrupts are important because they give the user better control over the computer. Without interrupts, a user may have to wait for a given application to have a higher priority over the CPU to be run. This ensures that the CPU will deal with the process immediatey.

⊙ Self-Quiz

Q1. What is the main purpose of the operating system?
Q2. What is the main aim of the operating system when a computer system is to be used in home?

INTERFACE OF OPERATING SYSTEMS

The way the user is interacting or using the operating system is called the *interface*. Operating systems normally have two types of interface, i.e., (i) a *character user interface* (CUI) and (ii) a *graphical user interface* (GUI).

Character User Interface (CUI)

Commands must be typed to give instructions to the computer hardware. The user needs to remember the commands for the proper use of computer system. To work on computer systems with a CUI operating system, users need to type the commands; otherwise the user will not be able to work.

Following are the characteristics of a CUI operating system:

- Tedious and slow (all commands must be typed)
- Less ease of use (compared to point and click sytems)

The Disk Operating System (DOS) is an example of a CUI-based OS.

Graphical User Interface (GUI)

The Graphical User Interface (GUI) is made up of mainly of five primary components: windows, icons, menu bars, pointing devices, and graphics. When a GUI is present, the user is interacting with one of these components at all times.

Windows: Multiple windows allow different information to be displayed simultaneously on the user's screen.

Icons: Icons are symbols that represent different types of information. On some systems, icons represent files; on others, icons represent processes.

Menus: Commands are selected from a menu rather than typed in a command language.

Pointing devices: A pointing device, such as a mouse, is used for selecting choices from a menu or indicating items of interest in a window.

Graphics: Graphical elements can be mixed with text on the same display. Some characteristics of a GUI include:

- There is no need to type the commands.
- Mainly consists of icons with pictures of commands.
- Commands are executed just by clicking on the icons.
- Easy to use.
- User may easily switch from one task window to another.
- User can work on many applications simultaneously.

Microsoft Windows is an example of a GUI-based OS.

Self-Quiz

Q3. Name an operating system that is CUI-based.

Q4. Write some advantages of a GUI-based operating system.

Where is the operating system present in the computer system?

The operating system is present in the hard disk of the computer. When the computer is switched on, at that time the OS from a hard disk is loaded into the main memory of the system. This process is known as the *booting* process. It is also known as *bootstrapping*.

Booting is a process to load the operating system from hard disk to main memory when the computer is switched on.

The address of the OS in the hard disk is present in the *Basic Input/Output System* (BIOS). The control unit reads the address (location) of the OS from the BIOS. The BIOS is software that is stored on a flash memory chip. In a computer system, the BIOS is embedded on the motherboard.

STEPS OF A BOOTING PROCESS

- Power button of computer is turned on.
- CPU pins are reset.
- The registers that are present in the CPU are set to a specific value.
- The control unit of the CPU jumps to the address (location) of BIOS.
- BIOS runs POST (Power-On Self Test) and other necessary checks.
- BIOS jumps to MBR (*Master Boot Record*). MBR is the register in a CPU that stores the data being transferred to and from the immediate access store.
- Boot loader program is present in master boot record.
- Boot loader loads operating system in main memory.

Self-Quiz

Q5. Write the full form of BIOS.

Q6. What do you mean by booting process?

OPERATING SYSTEM AND ITS FUNCTIONS

The operating system is actually complex software that is made up of various small modules. Each of its modules performs a pre-defined and specific task for which it has been designed. The various functions performed by the operating system are as follows:

- Prepares a list of all authorized users
- Maintains a list of all resources in the system
- Constructs a list of all processes running in the system
- Creates processes
- Initiates the execution of programs
- Provides controlled access of resources to the programs
- Performs CPU scheduling
- Allocates main memory to the processes
- Allocates resources to the processes
- Provides inter-process communication
- Provides security and protection

Did you know?	The first operating system on microcomputer platform was Control Program for Microcomputers (CP/M), developed for Intel 8080 in 1974 by Gary Kindall.

CLASSIFICATION OF AN OPERATING SYSTEM

CPU time is crucial. The CPU should not be idle at any time. Its time is considered to be very expensive. Various types of OS are designed to maximize the CPU utilization, which is considered as the primary goal. Almost all operating systems consist of similar components. They perform very similar functions. The *methods and procedures* used to perform these functions differ from system to system. Based on the ways these functions are performed in an OS, they are classified with many types.

- Single user system
- Batch operating system
- Interactive operating system
- Multi-programmed system
- Time sharing operating system
- Real time system

- Multiprocessor operating system
- Distributed operating system
- Network operating system
- Multiuser operating system
- Multithreaded operating system

Let's discuss each of them.

Single User System

Such systems are also known as early systems. They were developed in the early 1950s. Punch cards were used to store the programs. A significant amount of setup time is required. I/O devices are extremely slow. CPU utilization is lower, but such type of computer system was considered to be secure.

Batch Operating System

To reduce the setup time and to maximize the CPU utilization, the *batch system* was developed. The operating system in which programs are analyzed and categorizes them in batches of similar nature is called a batch OS. The programs present in each batch are of similar nature. The requirements for their execution are similar. The programs present in each batch are executed one after another. In this OS, the programs are submitted to the CPU for their processing and they are executed. In this type of system the user cannot interact with the program while it is running. Such systems are appropriate for executing large jobs that need no user interaction in producing the output. The user simply submits the job and returns later for its output. The process of a batch operating system is shown in Figure 1.4.

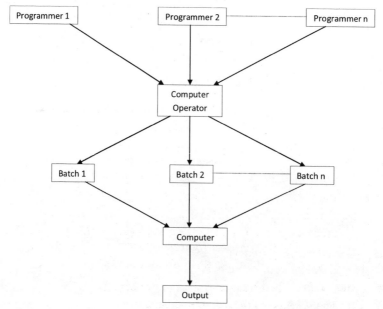

FIGURE 1.4. Batch Operating System

Such a system was developed in the early 1960s. In batch systems, the setup time of programs is diminished (reduced) by grouping similar type of programs. Similar programs are put in a batch. The programs were written on *punch cards* which is a mechanical device. They are submitted to the CPU for execution. While the programs are being executed, there is no interaction with the user. When all of the processes present in one batch are executed, then the next batch is set up. In transition from one batch to another, the CPU sits idle. As explained earlier, CPU time is crucial. The main purpose of any OS is to have maximum utilization of the CPU. To overcome this problem, a batch system uses *automatic job sequencing*.

Automatic Job Sequencing

Automatic job sequencing allows control from one batch to another to be transferred automatically. As soon as the programs of one batch finishes, control is transferred to the programs present in another batch. This process continues, until all of the programs present in each batch are finished. To implement automatic job sequencing, a program is used. This program is called a *resident monitor*.

Resident Monitor

A resident monitor is a program. It is also known as *job sequencer*. It is present in the main memory of the computer. With the help of this program, there is no idle time between executions.

Some disadvantages of a batch operating system follow:

■ There is a lack of interaction between the user and programs.
■ As there are speed mismatches between mechanical devices and the CPU, the CPU is often idle.
■ It is difficult to provide the desired priority to the different programs.

Examples of batch operating systems are (i) payroll systems, (ii) database updating systems, etc.

(◉) Self-Quiz

Q7. Name some application areas where batch operating systems could be useful.

Q8. Write some advantages of a batch OS.

Interactive System

An operating system in which the user can directly interact with the operating system is called an *interactive system*.

Features of interactive operating systems follow:

■ Such a system requires a proper user interface.
■ They usually process the data immediately.
■ They could be either command line interface or a graphical user interface.

■ In such a system, there are two-way communications between user and computer.

Examples of interactive operating systems are (i) ATM machines, (ii) spread- sheet applications etc.

Multiprogrammed System

There are two types of programs, *CPU-bound programs* and input-output *(I/O)-bound programs*. CPU-bound programs require more processing time. These programs use large amounts of computation. Performing computations or calculations, more CPU time is needed. I/O-bound programs are interactive programs. They need more input and output devices to finish their execution. This means such programs require more of devices rather than processing time (CPU time).

When a program is executed it may have some I/O needs. To fulfill the I/O need the program is switched from the CPU. To complete the need they are transferred to I/O devices. While the I/O need of a program is fulfilled, the CPU is idle. As explained earlier, CPU time is crucial. Hence some mechanism is needed so the CPU is busy at all the time. That is why the concept of *multiprogramming* came in existence.

The systems in which many programs are placed into the main memory of the CPU whenever the CPU is idle are called multiprogrammed systems. An example is shown in Figure 1.5.

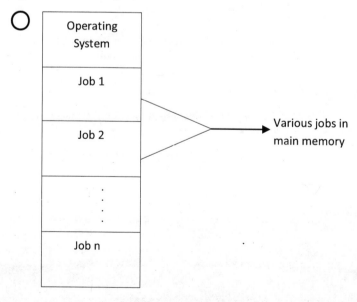

FIGURE 1.5. Multiprogramming Operating System

In multiprogrammed systems many programs are placed into the main memory. Whenever any I/O need appears in a program, during that time, the CPU is idle. During that time, another program is automatically selected and transferred to the CPU for its execution. When the I/O need of the first program finishes, then it is transferred back to the CPU. The CPU is switched from one program to another. Hence CPU is busy almost all the time. Whenever any I/O needs appear in a program, the CPU is assigned with another program, to keep it busy. Hence this way CPU is handling many programs at the same time – a multiprogramming system.

Features required for implementing a multiprogrammed system

- **Main memory management:** Many programs are to be kept in the main memory. Space in the main memory is limited. All programs cannot be kept in it at the same time. Higher priority programs should get space in main memory. Thus, main memory management is required so that all programs get space in it with time.
- **CPU scheduling:** There are many programs present in main memory that are ready for execution. For the execution of programs, they are to be placed in the CPU and all the programs get executed in time. To select the programs from the main memory and to transfer them to the CPU for their execution is called *CPU scheduling*.
- **Device management:** The resources like input and output devices, that are present in a computer system, are limited. In multiprogramming, many programs are placed in the system. All of the programs have to share the same resources. These resources are to be managed and used by the programs in a controlled manner. This is called *device management*.
- **Minimum switching time:** Whenever an I/O need appears in a program, The CPU switches to another program. Before switching to another program, the state of the previous program is to be stored. After that, the new program is loaded into the CPU. The time required in switching from one program to another is called *switching time*. This time should be minimal because the CPU is idle.

(•) Self-Quiz

Q9. When does the need for a multiprogramming OS arise?
Q10. Write some advantages of multiprogramming.

Time Sharing System

The system in which many users and programs share the CPU time is called a *time sharing system*. This type of OS uses the concept of multiprogramming and CPU scheduling. Many programs are kept in the main memory and hard disk. The CPU is multiplexed between various programs. The CPU's time is shared among various users or programs. This division of CPU time in small portions is called a *time slot*. Each program is executed in a particular time slot. When the time slot has expired, the CPU is switched to another program. The time slot is so small, that users feel that they are the sole user of the computer system during that period. To divide the CPU time in small portions and providing each portion to the program is called *time sharing*. It is also known as *time slicing*.

In Figure 1.6, user 4 is executing his program whereas user 1, user 2, user 3, and users 5, 6, 7, 8 are in a waiting mode. As soon as the time slice of user 4 finishes, the control will move to the next user, i.e., user 5 in the given example. The process continues in this manner until all programs of the users are executed.

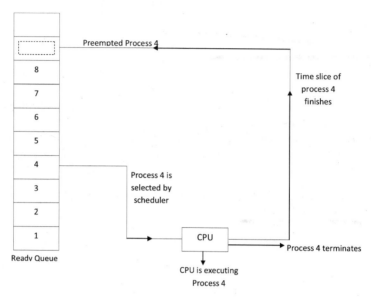

FIGURE 1.6. Time sharing operating system

Such a system allows many users to use or share the computer. In this system the programs are swapped between memory and disk. The time slot and switching time from one user to another or from one program to another is very fast; thus the user feels that he has the computer system at their own console. These systems are more complex as compared to multiprogramming systems because they use the concept of multiprogramming in conjunction with the concept of time slicing. This system is also known as *multitasking* or a *time slicing* system. The CPU time slot is also known as a time slice, quantum, and time chunk.

The advantages of a time sharing operating system are as follows:

- Response time is quick.
- As many users and programs are sharing the system, there is no duplication of software.
- The CPU is handling many programs at the same time which reduces the idle time of the CPU.

Disadvantages of a time sharing operating system are as follows:

- Less reliability.
- Security and integrity of user programs and data is lower.
- Data communication between programs must be handled.

An example of a time sharing system is UNIX.

Self-Quiz

Q11. What is the difference between a multiprogramming OS and a time sharing OS?

Q12. Which OS is more complex, time sharing or multiprogramming?

Real Time System

A real time application is a program that responds to activities in an external system within a maximum timeframe, which is determined by the system. *Real time systems* are those in which the operation on a process is to be finished within a designated time limit. In this system, processing must be done within the defined time constraints, or otherwise the system will fail. A real time system is considered to function effectively if it gives the correct results within the defined time limit. This system is in contrast with time sharing systems and batch systems.

There are two types of real time systems, i.e., *hard real time* and *soft real time*. Let's discuss each one of them.

Hard Real Time

The critical task must be completed in a defined time period. In these systems secondary storage of any type is not present.

Why is designing a hard real time system difficult?

- Many advantageous features that are normally present in other type of OS are absent in this system.
- The advantageous features like user friendliness makes the user separate from the hardware.
- The purpose of hard real time is to finish processing and produce output in a restricted time.
- In these systems virtual memory is absent. They cannot support a time sharing operation

Soft Real Time

This system is less restrictive as compared to a hard real time system. In this system a critical real time task gets priority over other tasks and the priority

TABLE 1.1. Examples of Real Time Systems

S. no.	Hard real time system	Soft real time system
1	Air traffic control	Traffic lights system
2	Industrial plant control	Multimedia applications
3	Missile guidance	Reservation systems
4	Nuclear power plant control	Banking systems
5	Robotics	Home appliance controllers

will remain until that task is completed. This system can be mixed with other types of systems

As these systems lack the deadline and fixed time constraint support, they cannot be used in critical applications like missile guidance, air traffic control, industrial controls, etc.

(◉) Self-Quiz

Q13. Why are hard real time systems difficult to design?

Q14. In which type of applications is a soft real time OS desirable?

Multiprocessor System

A computer system in which two or more CPUs work in communication with each other is called a *multiprocessor system*. They are also known as *parallel systems*. Following are some features of multiprocessor system:

■ Processor units can communicate with each other.

■ Processor units share memory and clock signals.

■ Processors communicate with each other with the help of shared memory and inter process communication (explained later in detail).

■ Processor units share common memory area.

■ The processor units might use different operating systems.

Advantages of multiprocessor systems are as follows:

■ **High throughput:** The total number of programs finished in unit time is called *throughput*. As more than one processor unit is embedded, more work is done in less time.

■ **Very economical:** As processor units are embedded in a single computer system, so they share the resources like I/O devices, clock, and bus. Hence it is economical because units share the devices.

■ **High reliability:** As more than one processor unit is working together, reliability is high.

■ **Fault tolerance:** If one processor becomes faulty for any reason, its work can be handled by the other processor.

■ **Graceful degradation:** When a processor unit becomes faulty and its work is shifted to the other processor, throughput may be lower. This means the throughput degrades but the whole system keeps working. It keeps on producing output although it may be slow. This is called *graceful degradation*.

Examples of multiprocessor systems are UNIX and OS/2.

(◉) Self-Quiz

Q15. What is the difference between multiprogramming and multiprocessing?

Q16. Why is a multiprocessor OS more complex as compared to a multiprogrammed OS?

Multiuser System

The system in which multiple users from different terminals or computers are made to access a system with one operating system is called a *multiuser operating system.*

The features of multiuser systems are as follows:

- Such systems are very complicated.
- Many users are allowed to access a single system; user management is very important.
- Multiple users are allowed to use the resources of that single system.
- Such systems normally run on mainframes.
- If they fail, then they affect all the users.
- Primarily used in offices where common resources are to be shared among users.
- Such an OS supports many users at the same time.

Examples of multiuser systems are Linux, UNIX, Virtual Memory System (VMS), Xinix.

Distributed Operating System

In this system, several physical processors are connected with each other through communication media. These processors can communicate with each other in a network. They may be geographically separated. The computation of large applications can be divided into small modules. Each module is distributed to a different processor. The processor units will perform the computation and the result is returned to the main processor. The output from each different processor is processed through one system to produce the overall output of the main application.

The features of *distributed operating systems* are as follows:

- They are also known as *loosely coupled systems.*
- In this system, each processor has its own memory.
- The processors could be of different configurations.
- Some processors might have higher processing power than others.
- They need Local Area Network (LAN) or Wide Area Network (WAN).
- Processors communicate with each other through communication media.
- Communication media could be wired or wireless.
- Wired media could be high-speed buses, telephone lines, fiber-optics etc.
- Wireless media uses satellites that also allow network connections for mobile users.

Advantages of Distributed Systems

- **Resource sharing:** As processor units are connected with each other, they can share the resources.
- **High speedup:** As work is divided in different modules and distributed among processors, large applications get executed in less time.

- ■ **Load sharing:** In place of executing big applications on a single processor, it is distributed among various processors. Thus, the load of one processor is shared among various units.
- ■ **Communication speedup:** The processor units are able to exchange information through communication lines.
- ■ **Fault tolerance:** If one unit fails in a distributed system, other processor units continue their operations. The system does not stop.

(◉) **Self-Quiz**

Q17. What do you mean by a distributed OS?
Q18. Find some disadvantages of a distributed OS.

Network Operating System

The systems in which various computers are connected with each other in a network so that they share resources and information like files, printer, fax machine, etc., is called a *network operating system.*

In this system the operating system runs on a high configuration system called a server. The server has the capability to manage information, users, applications, and networking functions.

Examples of network operating systems are Microsoft Windows Server 2003, Microsoft Windows Server 2008, UNIX, Linux, Mac OS X, Novell NetWare, BSD.

Advantages of network operating systems follow:

- ■ The centralized server in this system is stable.
- ■ Security is managed by servers.
- ■ Remote access to servers is possible from different locations.

Disadvantages of network operating systems are as follows:

- ■ High cost of the server.
- ■ Dependency on a central location for most operations.
- ■ Constant updates are needed.

(◉) **Self-Quiz**

Q19. Compare a network OS and a distributed OS.
Q20. Name some network operating systems.

SUMMARY

An interface between a user and the computer hardware is known as an *operating system.* The operating systems are designed with the aim of achieving two main goals: convenience to the user and the effective use of the computer hardware. The operating system manages the resources of the system including the CPU, memory, and I/O devices. Although all operating systems have similar components and perform similar functions, they are classified into different categories, depending on how these functions are performed. The main classification consists of batch, multiprogramming, time sharing, real time, multiprocessing, distributed, and network operating systems.

POINTS TO REMEMBER

- An interface between a user and computer hardware is known as an operating system.
- The OS works as an intermediate between computer hardware and user.
- It provides an environment for the user so he can run programs efficiently on the system.
- The operating system is software containing a group of programs performing various functions.
- The OS is also known as resource manager. It is called a resource manager because it manages and provides controlled access of resources to the various users and programs.
- The main resources of any computer system are the CPU, memory, and I/O devices.
- The programs in a system share all the resources of the system.
- All operating systems have similar components. The main components of any computer system are the hardware, the operating system, the application programs, and the users.
- Operating systems are mainly classified into batch, multiprogramming, time sharing, real time, multiprocessing, distributed, and network operating systems.
- In batch operating systems, the programs of similar nature and requirements are grouped together to make a batch.
- Placing many programs in main memory and putting a program to CPU whenever the CPU is idle is called multiprogramming.
- In a time sharing system, CPU time is divided in small portions.
- The CPU time is divided into small portions called time slices.
- In real time systems, the application needs to finish its execution within strict time constraints.
- There are two types of real time systems: soft real time and hard real time.
- In multiprocessing systems, there is more than one processor unit embedded on the same motherboard.
- In a distributed operating system, the geographically separate computer systems are connected with each other through communication links.

REVIEW QUESTIONS

Q1. **What is a real time operating system? What is the difference between hard real time and soft real time operating systems?**

Q2. **List the differences between a multiprogramming and a multiprocessing OS.**

Q3. Discuss essential properties of a time sharing operating system, a real time operating system, and a distributed operating system.

Q4. What do you mean by distributed systems? Discuss the advantages.

Q5. What are the main features of the following operating systems?
   i. Batch processing operating system;
   ii. Multiprogrammed operating system.

Q6. List the advantages of a batch processing system.

Q7. What are the main features of a real time operating system?

Q8. List the various steps of the booting process.

Q9. What do you mean by a boot strap loader program?

CHAPTER 2

OPERATING SYSTEM ARCHITECTURE

Purpose of this Chapter

The aim of this chapter is to explain the architecture of the operating system. Various ways in which an operating system can be designed are explained here. Further it reviews the concept of operating system functions, which are provided by various system calls. This chapter also focuses on various components of the operating system and its structure.

INTRODUCTION

The *operating system architecture* means the order in which certain hardware processes are carried out by the operating system, e.g., device management using device drivers, process management using processes and threads, inter-process communication, memory management, and file systems. The *kernel* is the core of an operating system. The prime function of the kernel is to run programs and provide secure access to the machine's hardware. Because there are many programs and resources are limited, the kernel also decides when and how long a program should run. This process is known as *scheduling*. The term scheduling represents the method by which threads, processes, or data flows are given access to system resources like processor time and communications bandwidth. This is performed to load balance and share system resources effectively.

ARCHITECTURE

Architecture defines the fundamental structure of an operating system. It defines the interconnection between the various system components. The various components of the system are hardware, operating system, application programs, and programming languages. They are shown in Figure 2.1.

FIGURE 2.1. Various system components

An operating system is designed using following different architectures as shown in Figure 2.2.

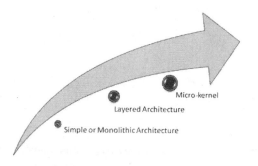

FIGURE 2.2. Different Architectures of an Operating System

■ Simple or monolithic architecture
■ Layered architecture
■ Microkernel architecture

Let us discuss each of the operating system architectures in brief.

Simple or Monolithic Architecture

Monolithic architecture has a small, simple structure, and hence is also called *simple architecture*. It only consists of a single layer, responsible for all the functions of the operating system.

■ The OS is written in the form of collection of procedure, so whenever the need arises, one process will call to another.
■ In terms of information hiding in a monolithic architecture, every procedure is visible to others and therefore in a monolithic system it is possible some structure.

Examples of a monolithic architecture system are the Microsoft Disk Operating System (MS-DOS) and UNIX

In a monolithic system, all the services are provided by the OS; for example system calls are requested by placing the parameter or arguments in defined places like the register or on a stack where they are executed by a TRAP instruction which refers to a supervisor call or kernel call(See Figure 2.3).

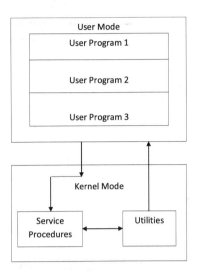

FIGURE 2.3. The Monolithic Architecture

Let's discuss TRAP instruction. The *TRAP instruction* command is used in assembly language. This command consists of a set of basic service routines that are used to simplify operations. Basically this instruction command is used by a program to transfer control to the operating system.

Two main components of a monolithic operating system are the kernel and the system program as explained below.

Kernel

The control module of the OS is the kernel. It is loaded initially at the time of the system boot and remains there until the system is switched off. Thus, it should be coded as small as possible while providing all of the essential services required by the remaining components of the OS and applications.

The kernel directly interacts with the hardware and schedules the execution of various tasks. For example, in multiple processor operating systems, the kernel chooses the processor and the program to be chosen for execution. The kernel performs low level functions, such as reading the input from the keyboard and displaying the output on the monitor. In addition to these low level functions, the kernel also performs other important functions, such as memory management, scheduling, and system accounting of various processes. The kernel is also called the *real time executive.*

Functions of the Kernel

- Memory management
- Process and task management
- Disk management
- CPU scheduling

System Program: The system defines the application program interface for UNIX where the set of system programs defines the user interface. The framework defined by the interface and the programmer must be supported by the kernel.

Advantage of a simple or monolithic architecture:

- Simple structural design

Disadvantages of simple or monolithic architecture follow:

- Lacks the ability to hide information
- Difficult to modify, as the entire OS has to be redesigned
- Failure of a single program crashes the entire system

Layered Architecture

Layered architecture was designed to reduce the limitation of a monolithic architecture. It categorizes the operating system in different layers, which communicates using standard function call as shown in Figure 2.4.

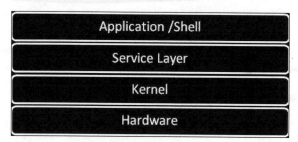

FIGURE 2.4. The Layered Architecture

Following are the features of the layered architecture:

- Layered architecture implements the concept of *information hiding* by providing the encapsulation of data in one layer and providing the functionalities and services only to the upper layer.
- By decomposing into several layers, each higher level layer uses the functions (operation) and services of its lower layers and thus maintains the modularity method.
- Easier to modify the layers. If an error appears in a particular layer, then only that layer is modified.

Figure 2.5 illustrates the communication process in layered architecture. The communication process is segmented into the following procedures:

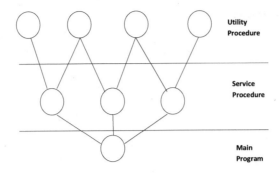

FIGURE 2.5. Communication in Layered architecture

■ A main procedure (program) requesting the services.
■ A set of service procedures that carry out system calls.
■ A set of utility procedures supporting the system calls.

Advantages of layered architecture follow:

■ Implements the concept of information hiding
■ Maintains a modular approach
■ Simplified debugging and system verification. If modification is required in some layer, then only that layer is redefined without troubling other layers.

Disadvantages of layered architecture follow:

■ Requires careful planning for designing the layers, as a higher layer can use the functionalities of only the lower layers.
■ Each layer adds an operating cost to its higher layer and takes a longer time to complete

Micro-kernel Architecture

Micro-kernel architecture is formed by removing all non-essential components from the kernel. Implementing separate levels for system and user level programs results in a small kernel called a *micro-kernel*. The prime function of micro-kernel architecture is to provide communication between the client program and other services that are running in the user space.

This feature is implemented by a message passing mechanism, where a client program required to access a file must interact with the file server, and there is no direct interaction between the client program and the particular services.

Advantages of a micro-kernel architecture:

■ Because more services are running as "user" processes than kernel processes, it provide more security and reliability to the system(See Figure 2.6).
■ If particular user services fail, there is no effect on the operating system.
■ Lower level services are only implemented within the kernel.
■ Its adaptability for use in a distributed system is another advantage of this architecture. If a client program wants to access a file from a server, then

there is no need for the client to know where it is handled locally on their own machine. In both cases, the client only sends the request and subsequently gets a reply.

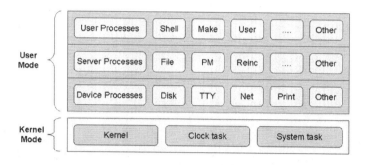

FIGURE 2.6. Micro-kernel Architecture

Self-Quiz

Q1. List the name of four services provided by an operating system.

Q2. What is a kernel? Explain the various operations performed by the kernel.

Q3. What is the main advantage of a layered architecture?

Q4. "Absence of structure or there is no structure" illustrates which operating system?

Q5. What are the needs of a micro-kernel architecture of an operating system? Explain the architecture in brief?

Q6. Give a comparison among different architectures of operating systems.

Q7. Explain the difference between the micro-kernel and the macro-kernel architecture.

OPERATING SYSTEM COMPONENTS

The environment provided by the operating system to run the program is constructed by partitioning it into small components. The components of operating system are shown in Figure 2.7.

Process Management	Main Memory Management	File Storage Management	Input - Output Management	Networking	Protection	Command Interpreter

FIGURE 2.7. Operating System Components

Let's discuss each of the components:

1. **Process Management:** To execute a whole program, instead of loading the whole program in the main memory, the CPU fetches and executes one instruction of a process after another.

   Optimum:

   ■ Whenever the CPU fetches data or instructions, it is always from the main memory.

   Responsibilities:

   ■ Creation, loading, execution, suspension, resuming and termination of the process.
   ■ Scheduling of the process, i.e., to switch system between multiple processes in the main memory.
   ■ Offers mechanism of communication between processes (send or receive data to or from processes).

   Process: It is a fraction of the program that is loaded in the main memory.

   ■ Provides synchronization (control concurrent access) to share data to keep shared data consistent.
   ■ Prevents or avoids deadlock situation by properly allocating or de-allocating resources.

2. **Main Memory Management:** To execute a process, it must be loaded into main memory. By increasing the "Hit Ratio," this component increases system performance. Thus, the memory utilization is maximized.

   Responsibilities:

   ■ This component keeps track of particular memory sections utilized by processes.
   ■ As per the requirements, it allocates or de-allocates memory to the processes.

3. **File and Storage Management:** Most of the data is stored in the secondary memory. Therefore access of secondary storage must be efficient by performance level and convenient to program input/output functions at the application level.

   Responsibilities:

   ■ Creation, manipulation and deletion of files or directories are the major responsibility of file management component.
   ■ Allocation, de-allocation, defragging blocks, marking of bad blocks and scheduling of several input/output requests for performance optimization are some prime responsibilities of storage management component.

4. **Input / Output Management:** It keeps the details away from applications to ensure proper utilization of the devices, avoid errors, and offers end users

an efficient and convenient programming environment. The operating system provides an abstract level of hardware devices.

Responsibilities:

- Hides the needless details of hardware devices.
- By using cache, buffer and simultaneous peripheral operations online (*spooling*), it manages main memory for devices.
- It is also responsible for providing and maintaining device driver interfaces.

5. **Networking:** In a distributed system, each processor has its own local memory and clock. There is a communication line known as a network through which processors communicate with each other. This network may be partially or fully connected. The operating system simplifies network access in the form of file access along with the details of networking of various network interfaces.

6. **Protection:** This component provides protection from erroneous programs and malicious programs to hardware resources, kernel code, processes, files, and data.

7. **Command Interpreter:** This component permits the end user to interact with the operating system and offer the user a compatible environment for programming.

(◉) Self-Quiz

Q8. What is the role of process management?

Q9. Describe five components of the operating system along with their responsibilities.

Operating System Services

Figure 2.8 depicts the main services provided by the operating system for the end user.

1. **User Interface:** Users can issue commands to the system by means of user interfaces. These may be from command line interfaces, Graphical User Interface (GUI) interfaces or a batch command system.

2. **Input / Output Operation:** Transferring of data to and from input/output devices like keyboard, monitor, and storage devices are the prime functions of the operating system.

3. **Program execution:** The operating system must be capable of loading programs into the *Random Access Memory* (RAM), running and termination (normally or abruptly).

4. **Communication:** Inter-process communication takes place either between processes running on separate processors on separate machines or on the same processor. It can be implemented by either message passing or shared memory (or some systems are provided by both).

5. **File System Manipulation:** Along with the storage of raw data, the operating system is also responsible for maintaining the directory and subdirectory

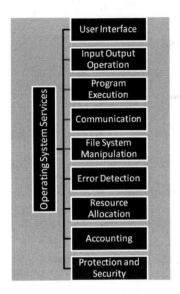

FIGURE 2.8. Operating System Services

structure, mapping of file names to the memory blocks, and for navigating and using the file system.

6. **Error Detection:** Both software and hardware errors must be properly detected and handled with minimum harmful impact. Some systems may include errors e.g., *Redundant Array of Inexpensive Disks (RAID)*, and other redundant systems for error avoidance or recovery systems.

7. **Resource Allocation:** When multiple processes are running at the same moment, resources must be allocated to each. Different resources like the main memory, CPU cycles, and file storage are managed by the operating system.

8. **Accounting:** Keep track of system activity to optimize future performance.

9. **Protection and Security:** Preventing harm to the resources and to the system as a whole, either by internal processes or malicious outsiders. By authentication, ownership and restricted access OS provides protection and security to the system.

Self-Quiz

Q10. Write the full form of RAID.

Q11. How are security and protection services managed by an operating system?

SYSTEM PROGRAM

Unlike an application program, the *system program* is a program that can be an operating *system*, a compiler, or a utility *program*, that controls aspects of the operation of the computer.

The following are some basic categories of system programs.

1. **Status Information:** Programs involved in asking time, date, free disk space, available memory, user numbers, or similar information from the operating system.
2. **File Management:** Programs responsible for the creation, deletion, renaming, copying, dumping, deleting listings and manipulating files, and how directories are managed.
3. **File Modification:** A program that includes text editors to create or modify the contents of stored files.
4. **Status Information:** Programs responsible for creating the virtual connection between the user process and different computer systems.
5. **Program Language Support:** Compilers, assemblers, and interpreters for programming languages (like C, C++, VB, and Java) are provided for the user with the operating system.
6. **Programming Loading and Execution:** When program compilation is complete, the next step is to load it into the memory for execution. The requirement of debugging the system lies in this phase. The system might provide loaders, linkers, and overlay loaders.
7. **System Utilities:** Common operations, such as the word processor, spreadsheets, database systems, Web browsers, and many others are performed by these programs.
8. **Communication:** Virtual connection between the user's process and different computer systems are created by this program.

SYSTEM CALL

The *system call* provides an interface between the operating system and its processes. By making a system call, the user program communicates with the OS and requests services from it. Every system call has a library procedure that the user program calls. This procedure puts the parameters of the system call into the register and then issues a *TRAP instruction*, which is a protected procedure call to start the operating system. This library procedure hides the details of the TRAP instruction and makes the system call appear as an ordinary procedure call.

System calls can be grouped roughly into five major categories as summarized in Table 2.1.

⊙ Self-Quiz

Q12. Compare and contrast system call with system program.
Q13. List various utility programs run on a personal computer.

SUMMARY

Architecture defines the fundamental structure of an operating system. The control module of the OS is the kernel. It is initially loaded at the time of the system boot and remains there until the system is switched off. The different architecture on which the OS is designed is either simple, monolithic, layered, or microkernel. By implementing separate levels for system and user level programs results in a small kernel called a microkernel. The operating system provides various services including the user interface, input/output operations, program execution, communication, file system manipulation, error detection, resource allocation, accounting, protection and security.

TABLE 2.1. List of System Call Categories

System Call Categories	Functions Performed
Process Control Load or execute program File Management	• Create or terminate a process • Abort the program • Get or set file attribute • Create/Open/Read/Write/Close/ Delete files
Device Management Request/Read/Write/Release devices Information Maintenance Communication	• Get or set device attribute • Logically attach or detach devices • Get or set system data • Get or set time or date • Get or set process/file/device attributes • Create or delete communication connection • Send or receive messages • Attach or detach remote devices

POINTS TO REMEMBER

- The kernel directly interacts with the hardware and schedules the execution of various tasks.
- The operating system simplifies the access of networks in the form of file access along with the details of networking of the various network interfaces.
- Along with the storage of raw data, the operating system is also responsible for maintaining directory / subdirectory structure, mapping of file names to the memory blocks, and for navigating and using the file system.
- Compilers, assemblers, and interpreters for programming languages (like C, C++, Visual Basic, Java) are provided to the user with the operating system.
- The system call provides an interface between the operating system and processes.

REVIEW QUESTIONS

Q1. What are the major functions of the operating system?
Q2. Explain the following in brief:
    (a) System Protection
    (b) System Components

Q3. What are the advantages of the layered approach to the design of an operating system?

Q4. Define the monolithic kernel and microkernel.

Q5. How is a system call made? How is a system call handled by the system? Choose suitable examples for your explanation. Con conem sunt. Lestis restibus.

CHAPTER 3

PROCESS OVERVIEW

Purpose of this Chapter

This chapter defines the concept of process. It explains various states of the process and different state transitions. It gives details of the process control block that keeps all the information related to a process. The chapter further explains the different types of schedulers and their uses.

INTRODUCTION

With the development of multiprogramming systems, multiple programs need to be loaded into the main memory for their execution. If many programs are to be kept in the main memory, then each program will require ample space. Thus, fewer numbers of programs can be placed into the main memory. This is why programs are divided into small chunks that are called the *compartmentalization of programs*, or the breaking of the programs into small portions. These small portions that are present in main memory are called *processes*. This aids in keeping portions of several programs in the main memory at the same time.

THE MEANING OF PROCESS

A *program is a group of instructions which, when executed, provides a meaningful output.*

The part of the program that is present in the main memory is called a *process*.

A program is present in the disk whereas a process is present in the main memory. Thus, a program is considered as a passive entity whereas a process is considered as an active entity. A process is a part of the program which is ready to be executed and is present in main memory. *There could be any number of processes that belong to a single program.*

As mentioned, dividing the program into processes helps in keeping many programs in the main memory simultaneously. As a result, the degree of multiprogramming increases. The degree of multiprogramming means the total number of processes that are present in the main memory at the same time. A process can also be considered as a set of sequential steps for performing a particular task. In other words, as explained above, a process is an instance of a program. For example in the Windows environment, if you edit two text files simultaneously in *Notepad*, then you are implementing two distinct instances of the same Notepad application. In an operating system these two instances are separate processes of the same application.

State of the Process

A process is a program in the execution phase. During its lifetime, a process goes through various states. The *state of the process* means the activity a process is currently executing. A process state is indicative of the activity it is currently performing. For example when a process is created, it is in a *new state*. When a process is being executed, it is in a *running state*. When it is waiting for some resource, then it is in *wait state*, and so on. Thus, the main states of the process are *new, ready, running, wait,* and *terminate*.

Process Transition

When a process goes from one state to another it is called a *process transition*.

The transition of the process depends upon the flow of the execution of the process. Although there are various states of a process, it is not necessary that the process undergoes all of the states during its lifetime. The main states are as follows:

- **New:** When a process is created, it is in the *new state*.
- **Ready:** After its creation, a process needs to be executed. For execution a process uses the central processing unit (CPU). In this condition a process is said to be in *ready state*.
- **Running:** When a process is in execution mode, it is called in a *running state*.
- **Waiting:** During execution, a process may need input/output devices. When a process needs devices, then at that time its state is changed to the *wait state* because it is waiting for devices to complete its execution.
- **Terminate:** When a process finishes its execution, it is said to be *terminated*. Then it is removed from the memory.

Various states of the process are shown in Figure 3.1.

A new process is admitted into a data structure called a *ready queue*. This ready queue is also known as a *ready pool* or *pool of executable processes*. A *queue* is a *data structure* that stores all the data items in a "first in/first out" manner. A queue has two ends, i.e., the front and the rear. A process is inserted into the ready queue from the rear end and the processes from the front end of the queue are sent for execution by the scheduler. In other words, the processes are inserted into the ready queue from where they are selected and sent to the CPU for their execution.

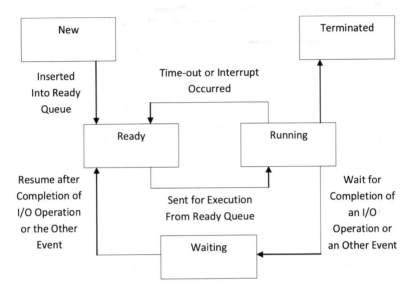

FIGURE 3.1. Transitions of the Process States

Each process is assigned a time slice for execution. A time slice is a very short period of time. It is also known as a *time quantum*. Duration or length of the time quantum varies from system to system. The CPU executes the process at the front end of the ready queue and the process is said to be in a running state. A *time out* occurs if the process does not finish its execution in that time slice. An *interrupt* is a request to the processor usually activated by a task that needs CPU attention. For example while executing a program, if an error occurs, then the program flow is interrupted and the error handling module will be executed. When a process is interrupted or a time out occurs, the running process is sent back to the ready queue.

When a process waits for the completion of an Input / Output task or the termination of another process, it is sent to the waiting queue and the process state is marked as waiting. After completion of the I/O task or termination of the other process, the waiting process is restored and placed back in the ready queue and the process state is marked ready. When execution of a process ends, the process state is marked as terminated and the operating system reclaims all the resources allocated to the process.

The process consists of several CPU and I/O related tasks. The CPU related tasks are performing arithmetic operations, such as addition, multiplication, and logical. Examples of I/O operations are reading data from a file and writing data to the file.

An I/O related task usually takes more time for completion than a CPU related task because the CPU is faster than I/O devices. As a result of this process, consisting of a sequence of both of these tasks, the CPU-related tasks may have to wait for the completion of the slower I/O related tasks. A process is said to be in a waiting stage if it is waiting for the completion of an input/output related task.

Almost all processes alternate between two states in a continuing cycle, a *CPU burst* of performing calculations, and an *I/O burst*, waiting for data transfer in or out of the system. In a process, the I/O and CPU bursts alternate. This means in a process a CPU burst occurs, then followed by another I/O burst, followed by another CPU burst, and so on.

A process is terminated after its successful completion. This is called the *normal exit* of the program. In certain cases, a process terminates or stops its execution prematurely. For example, when a process that kills another process or an error occurs that results in an abnormal termination.

⊙ Self-Quiz

Q1. Define process.
Q2. How is a process different from a program?
Q3. Name the various states of a process.
Q4. When does a process move from the running state to the waiting state?

PROCESS CONTROL BLOCK

The Group of memory locations where information related to a particular process is kept is called the *process control block (PCB)*. In other words, the data structure that stores the information about a process is called a process control block.

PCB is also known as a *task control block*. When a process is created, all of its information is kept in a PCB. The operating system keeps all the information of a process from its creation until its termination. When a process finishes, its process control block is released from the main memory. The various contents of the PCB are shown in the Figure 3.2.

The contents of PCB are discussed below.

- **Process ID**: Every process has an *identification number*. This identification number is used whenever a process is referred to. It is also used during inter-process communication (discussed later).
- **Process state**: During its lifetime, a process can be in one of its various states. The different states of the process are new, ready, waiting, running, and terminate.
- **Pointer**: The content of the *pointer* points to the next PCB in the scheduling list. It is used when the CPU needs to switch from one process to the next process.
- **Registers**: A *register* is a special, high-speed storage area within the CPU that is stored during interrupts, so linking is maintained and will occur

correctly when the process resumes. In this field of the PCB, it names the registers and values that a process is using.

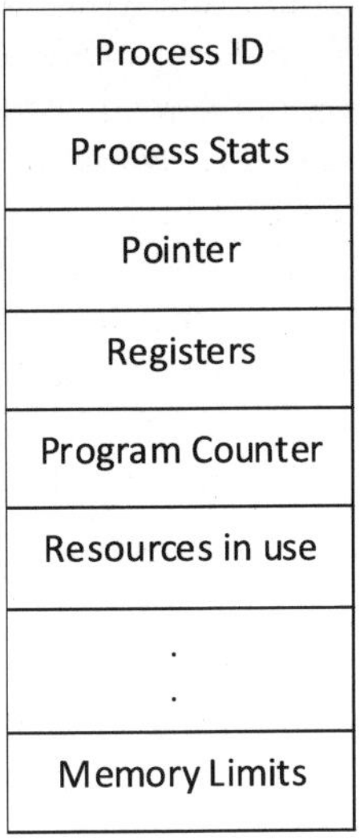

Process ID
Process Stats
Pointer
Registers
Program Counter
Resources in use
. .
Memory Limits

FIGURE 3.2. The Process Control Block

- **Program counter**: It indicates the address of the next instruction which is to be executed within a process.
- **List of open files**: This field includes all the files that are used in a process during its execution.
- **Resources in use**: It contains the resource names that are used by the process.
- **Memory limits**: This field gives the size of a process. It gives the starting address of the process in the memory.

In addition to these listed above, there could be other fields in a PCB like the priority of the process, its link to the parent process, link to the child process, its I/O status, etc. Each process has a priority, implemented in terms of numbers. The higher *priority processes* have precedence over the lower priority processes. The *priority field* is used for storing the priority of a process. A new process can be created from an existing process. The existing process is called the *parent* of the newly created process. The field link to the parent process stores the address of the process control block of the parent process in the main memory. The field link to the *child processes* stores the address of the process control blocks of the child processes in the main memory.

The PCB stores information regarding the programming environment of a process. The programming environment information includes the value of the registers, stack, and the program counter.

Registers hold the data to be processed, the results of the calculations or addresses pointing to the location of the desired data. A *program counter* is a register that stores the address of the next instruction to be executed. The width of the program counter is equal to the size required by the address of all possible instructions. *The stack* is used to store data in the *Last In First Out* (LIFO) order.

A process control block also stores information regarding the memory management such as the amount of memory units allocated to each process and the addresses of the memory chunks allotted. Information regarding I/O status (such as the I/O devices allocated to the process and pending I/O tasks) is also stored in the PCB.

The PCB stores file management information. Examples of the file management information stored in the PCB are the number of open files, the list of open files, and the access right of the open files, such as read-only, or read-write.

(◉) Self-Quiz

Q5. What is the main function of the process control block?
Q6. Is the process control block the same for all processes?
Q7. Where is the process control block stored?

SCHEDULING QUEUES

Several processes exist concurrently in the system. All processes have various needs. Some may need CPU for execution, whereas some others might need I/O devices. Because the resources are limited, some processes might need to wait. While waiting, according to the need, they are placed in different types of *queues*. When a process undergoes various states, it also stays in various queues. The main queues are discussed below.

- **Job Queue**: When a process is created and it enters in a system, it is placed in a *job queue*.
- **Ready Queue**: When a process is waiting for execution in the CPU, it is placed in a *ready queue*.
- **Device Queue:** When a process is waiting for the input/output devices for fulfilling its I/O needs, then it is placed in a *device queue*.

PROCESS SCHEDULERS

As explained above, a process might need to wait in one of the various queues. A process might move from one queue to another and from one state to another.

The part of the operating system that carries out this selection of the process is called the *scheduler*. Depending upon the place from where the scheduler is selecting the process, there are three types of schedulers.

1. Long Term Scheduler
2. Medium Term Scheduler
3. Short Term Scheduler

Let's discuss each one of them:

■ **Long Term Scheduler:** This scheduler is also known as the *job scheduler*. Initially all the processes are in the disk. From the disk the processes are selected and placed into the main memory. The scheduler which selects the processes from the disk and places them into the main memory is known as the *long term scheduler*.

■ **Medium Term Scheduler:** Sometimes processes need to be removed from the main memory to improve the process mix. Some processes are selected and removed before their termination. The scheduler which selects process from main memory and swaps them in disk is called the *medium term scheduler*.

■ **Short Term Scheduler:** This scheduler is also known as the CPU Scheduler. It selects the process from a ready queue and places it into the CPU for its execution. The scheduler that selects a process from the ready queue and places it into the CPU is called the *short term scheduler*.

The working of various types of schedulers is shown in Figure 3.3.

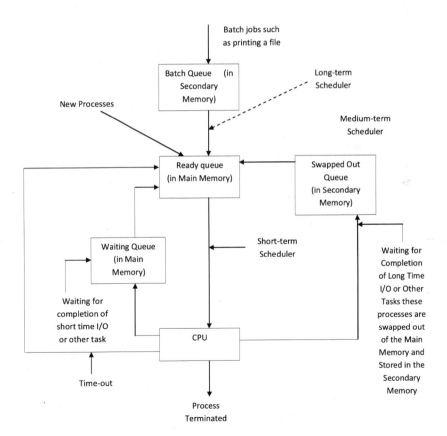

FIGURE 3.3. Roles of Various Types of Schedulers

Purpose of the Schedulers

Scheduling criteria varies from one schedule to another. The common aims for all schedulers are as follows:

- Implementing a scheduling plan such that all processes are given equal opportunity to execute.
- Creating a good mix of I/O and CPU-oriented processes.
- Utilizing and optimizing the system resources, such as the CPU.
- The term *CPU utilization* is defined as the percentage of time for which a CPU is busy executing instructions of a process, while minimizing the turnaround time. The amount of time to accomplish the execution of a process is known as the *turnaround time of the process*.
- Minimizing the waiting time. The amount of time a process waits in the ready queue to get a chance for execution is known as the *waiting time of the process*.

In a specific system, the goal of a scheduler may depend on the nature of the system. Examples of a specific goal of a scheduler in particular type of systems are:

- An automated teller machine (ATM) in an interactive system that is used for withdrawal of a sum of money. The user expects that a system should respond quickly. In other words, the response in an interactive system should be very short. *Response time* is defined as the period from the submission of a process to the OS for execution-to the appearance of the effect it initiated.
- In a real-time system, such as Internet banking, a scheduler focuses on security of data and prevents data loss.

⊙ Self-Quiz

Q8. **What do you mean by a scheduler?**
Q9. **What is the main purpose of a scheduler?**
Q10. **Write the differences between short term and long term schedulers.**

SUMMARY

A process is a program in an execution phase. During its lifetime, a process goes through various states. A process state is indicative of the activity it is currently performing. The primary states of a process are new, ready, running, wait, and terminate. The group of memory locations, where information related to a particular process is kept, is called a process control block. The part of the operating system that carries out this selection of the process is called a scheduler.

POINTS TO REMEMBER

- A process is a program in the execution phase.
- A program is present in the disk, whereas a process is present in the main memory.
- The state of the process means the activity a process is currently doing.
- The main states of a process are new, ready, running, wait, and terminate.
- The group of memory locations, where information related to a particular process is kept, is called a process control block.
- PCB is also known as a *Task Control Block.*
- The operating system keeps all the information of a process from its creation until its termination. When a process finishes, its process control block is released from the main memory.
- The part of the operating system that carries out this selection of the process is called a scheduler.
- The scheduler that selects the processes from the disk and places them into the main memory is known as the long term scheduler.
- The scheduler that selects a process from the main memory and swaps them into the disk is called the medium term scheduler.
- The scheduler that selects a process from the ready queue and places it into the CPU is called a short term scheduler.
- The short term scheduler is also known as a CPU scheduler.

REVIEW QUESTIONS

Q1. Draw the process state diagram and explain the state transaction.

Q2. Explain the Process Control Block (PCB). Discuss each component of the PCB.

Q3. Explain the need for the process control block.

Q4. Explain the term, *process.*

Q5. Differentiate between a process and a program.

Q6. Discuss the difference between short term, medium term, and long term schedulers.

Q7. Define different states of a process with a diagram. Explain the need for process suspension.

CHAPTER 4

THREADS

Purpose of the Chapter

In this chapter the concept of threads will be explained. A *thread* of execution is the smallest sequence of programming instructions that can be managed independently by a scheduler, in an operating system. This chapter explains multithreading, its need, and benefits. It further explains kernel level, user level, and hybrid level threads.

INTRODUCTION

A *thread* of execution is the smallest sequence of programming instructions that can be managed independently by a scheduler, in an operating system. It is also called a *Light Weighted Process* (LWP) with a reduced state. It is the smallest unit of the Central Processing Unit (CPU) utilization. A thread goes through different stages like new, ready, waiting, running, and termination- much like the process goes. By reducing the overhead of process switching, it also represents a software approach for improving performance of operating system.

MOTIVATION BEHIND THE CONCEPT OF THREAD

A process with multiple threads can perform several tasks, simultaneously. In a process, threads are typically used for dividing a task into a number of subtasks, which can run independently and can share data as well.

A common practice of designing a *multithreaded program* is to implement a main thread or primary thread of execution, which forks into a set of other threads.

WHY IS MULTITHREADING USED?

Consider the example of implementing multiple threads in the context of a word processing program. The word processor saves the current open document every three minutes. There can be multiple threads to implement various tasks in the word processor. While one thread is receiving input text from the user and another thread checks the grammar and spelling, the third thread performs the backup operation.

If the above example ran on a traditional single thread process, it would be able to execute only one process and the other process would have to wait for their turns.

Take another example where the task in a Web browser is divided into multiple threads: downloading images, downloading text, displaying images and displaying text. Downloading an image takes more time than downloading text. While one thread is busy downloading the image, another thread can utilize this time by displaying the text. Benefits of multithreading are shown in Figure 4.1.

Let's discuss each one of them.

1. **Responsiveness:** Multithreading allows us to run a program even if a part is blocked or is performing a lengthy operation. It reduces the overhead; thus it increases the responsiveness of user.

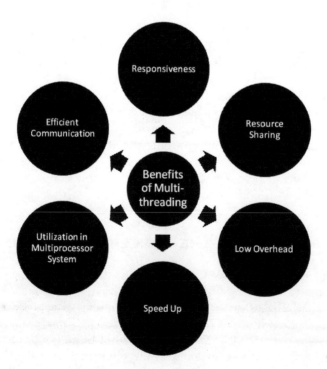

FIGURE 4.1. Benefits of Multithreading

2. **Resource Sharing:** The thread shares the memory and the resources of the process, allowing an application to have several different threads of activity all within the same address space.

3. **Low Overhead:** This thread state is the only state of computation. Other states like the resource allocation state and the communication state are not the part of the thread state and hence reduce overhead.

4. **Speed Up:** Concurrency can be realized by creating multiple threads. This speeds up the execution of an application in both the uniprocessor and the multiprocessor.

5. **Utilization in Multiprocessor System:** A single thread process runs on one processor, regardless of processor availability. Multithreading in a multiprocessor machine increases concurrency. The benefits of multithreading is efficiently increased with multiprocessor architecture where each thread might run in parallel on a different processor.

   In a uniprocessor system, the CPU generally moves very quickly between each thread to create an illusion of parallelism, but in actuality, only one thread is running. Threads may run in concurrency, but are not parallel.

6. **Efficient Communication:** Through shared address space, threads of a process can communicate with one other; thus avoiding the overhead of a system call for communication.

Self-Quiz

Q1. Define threads.
Q2. Why is threading desired?
Q3. What is multithreading?
Q4. List four benefits of multithreading.

IMPLEMENTATION OF THREADS

The three methods of implementing threads are:

1. *Kernel-Level Thread*
2. *User-Level thread*
3. *Hybrid Thread*

Kernel-Level Thread

The *kernel* is responsible for implementing a *kernel-level thread*. Therefore creation and termination of the kernel level thread and checking its status are done by system calls.

When a create-thread system call is invoked by some process, the kernel assigns an ID to it, and a *Thread Control Block* (TCB) is allocated to it. The TCB holds the pointer to the Process Control Block (PCB) of the process. When certain events occur immediately, the kernel saves the CPU state of the interrupted thread in its TCB. After handling the event, the scheduler considers a TCB of all threads and selects one ready thread. The dispatcher uses a PCB pointer in its TCB to check whether the selected thread belongs to a different process than the interrupted thread. If this happens, then it saves the context of the process to which the interrupted thread belongs, and loads the context of the process to which the selected thread belongs. After this, it dispatches the selected thread. Kernel-level threading is shown in Figure 4.2.

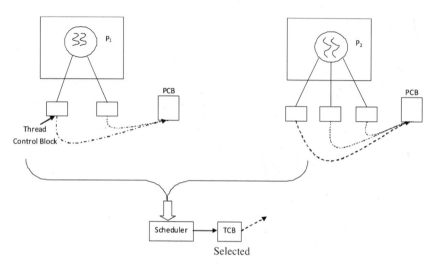

FIGURE 4.2. Kernel-Level Thread

Advantages

■ Similar to a process, except that it has a smaller amount of stated information.

■ Programming for a thread is similar to the programming for a process.

■ In a multiprocessor system, kernel-level threads provide parallelism, i.e., several threads belong to a process and can be scheduled simultaneously.

Disadvantage

■ Switching between threads is performed by the kernel as result event-handling arises, even if the interrupted thread and selected thread belong to the same process. Thus it bears switching overhead.

User-Level Thread

The *user-level thread* is implemented by a thread library. The library sets up the thread implementation without using a kernel and interweaves the operation of threads in the process. As the kernel sees only the process, it is unaware of the presence of user-level threads in it. While the dispatcher dispatches the process, the scheduler considers PCBs and selects a ready process. The user-level thread is shown in Figure 4.3.

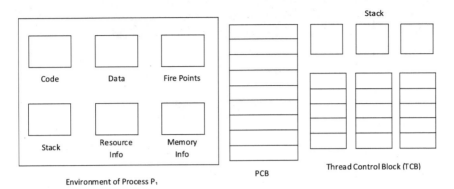

FIGURE 4.3. User-Level Thread

Advantage
- Thread switching overhead is similar to the kernel-level thread because thread scheduling is implemented by the thread library.

Disadvantages
- The kernel does not differentiate between the thread and a process.
- Because the thread library schedules the thread in the process, and the kernel schedules a process, *only one thread of a process is executed at one time*. Thus, the user-level thread does not support parallelism.

Hybrid Thread

Hybrid means mixture. The *hybrid thread model* has a kernel-level thread and a user-level thread and a method for associating the kernel-level thread with the user-level thread.

There are three methods in which the thread library creates a user-level thread in a process and associates them with a *Thread Control Block (TCB)*, and/or the kernel creates a kernel-level thread and associates them with a *Kernel Thread Control Block* (KTCB).

1. **Many-to-one association:** A single kernel-level thread is created in each process by the kernel, and all user-level threads created by the thread library are associated with this single kernel-level thread.

   Advantages of the many-to-one association are as follows:

   - User-level thread can be concurrent without being parallel.
   - Thread switching results in low overhead.

   A disadvantage of the many-to-one association is:

   - Blocking of a user-level thread leads to blocking of all threads in the process.

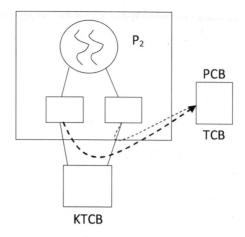

Many-to-One

FIGURE 4.4. Many-to-One Association

2. ***One-to-One Association:*** Each user-level thread is permanently mapped into a kernel-level thread.

   Advantages of the One-to-One association are:

   ■ Threads can operate in parallel to different CPUs of a multi-processor system.
   ■ Blocking of a user-level thread does not block others threads.

   A disadvantage of the One-to-One association is:

   ■ Switching between threads is performed at the kernel level and results in high overhead.

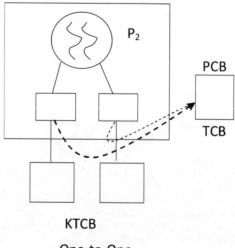

KTCB

One-to-One

FIGURE 4.5. One-to-One Association

3. *Many-to-Many*: User-level threads can be mapped to any kernel-level thread. Advantages of the Many-to-Many association are:

- Parallelism is possible.
- Switching is possible at the kernel level with low overhead.
- Blocking of the user-level thread does not block other threads as they are mapped with different kernel-level threads.

A disadvantage of the Many-to-Many association is:

- Requires complex implementation like the *Sun Solaris OS*.

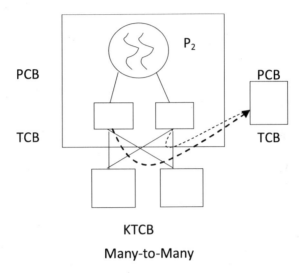

FIGURE 4.6. Many-to-Many Association

SUMMARY

A *thread* of execution is the smallest sequence of programming instructions that can be managed independently by a scheduler in an operating system. A thread goes through different stages like new, ready, waiting, running, and termination, similar to the process. The three methods of implementing threads that have been explained in the chapter are the kernel-level thread, the user-level thread and the hybrid thread.

POINTS TO REMEMBER

- A thread is a single, sequential flow of execution of the task belonging to a process.
- A thread goes through different stages like new, ready, waiting, running, and termination.
- The three methods of implementing threads explained in this chapter are: kernel-level thread, user-level thread, and the hybrid thread.

■ The user-level thread is implemented by the thread library.

■ A hybrid-thread model has a kernel-level thread, a user-level thread, and a method for associating the kernel-level thread with the user-level thread.

REVIEW QUESTIONS

Q1. Differentiate between a user-level thread and a kernel-level thread.

Q2. What is a thread cancellation?

Q3. List two differences between the hybrid and kernel-level threads.

Q4. What are the advantages of multithreading?

CHAPTER 5

CPU SCHEDULING

Purpose of the Chapter

This chapter defines the concept of CPU scheduling and the main considerations when choosing a particular scheduling algorithm. It further explains
various types of scheduling algorithms and their drawbacks. To provide better
clarity, each algorithm is explained with the help of examples.

INTRODUCTION

In a multiprogramming operating system, the method for switching the
CPU among multiple processes is called *CPU scheduling*. Scheduling is the
operation of selecting a process which is to be serviced by the CPU. A good
scheduling technique is very important in evaluating the performance of the
system. Various scheduling techniques exist that influence the performance of
the computer system.

WHAT IS CPU SCHEDULING?

The switching of the CPU from one process to another whenever any need
in a process arises (other than processing) is called *CPU scheduling*.

In a *multiprogramming operating system*, the method for switching the
CPU among multiple processes is called CPU scheduling.

SCHEDULING CRITERIA

There are several scheduling techniques. When selecting the preferred
method, there are different criteria. These criteria depend upon the particular
situation and environment required. The main points are as follows:

- **CPU utilization:** CPU time is expensive; thus it should be busy at all times. Ideally it is said that CPU utilization should be 100%, but practically it is difficult to achieve. During heavy loads, the CPU is busy almost 90% of the time and in lighter loads it is only active around 40% of the time.
- **Throughput:** The total number of processes that gets completed in unit time is called *throughput*. Throughput should be as high as possible.
- **Turnaround time:** The time span from submission of a process to the system, until its completion, is called *turnaround* time. Its value should be as low as possible.
- **Waiting time:** The time spent by a process in different queues, while waiting for CPU, is called *waiting time*. Its value can also be calculated by subtracting the execution time from the turnaround time of the process. Its value should be as low as possible.
- **Response time:** The time taken by a process in producing its first response after its submission is called *response time*.

While choosing any scheduling algorithm, CPU utilization throughput should be at a maximum, whereas turnaround time, waiting time, and response time should be at minimum.

Self-Quiz

Q1. Define CPU scheduling.
Q2. Define turnaround time and waiting time?

TYPES OF SCHEDULING

The CPU scheduler performs the task of scheduling and is a part of the operating system. There could be two types of scheduling algorithms, *preemptive or non-preemptive.*

Preemptive Scheduling

When the CPU switches from one process to another before its completion, then it is called *preemptive scheduling*. In preemptive scheduling, the CPU can preempt from a running process and switches to another. When this occurs, the process does not relieve the CPU voluntarily (by its own choice). There are many reasons why the CPU leaves one process and starts executing another. A few reasons follow:

- Some higher priority process arrives in the system. In such a situation, the execution of the low priority process is suspended and the high priority process is executed.
- An interrupt occurs in a process.
- A child process comes into a parent process.

Non-preemptive Scheduling

The CPU executes the process until it is terminated or until any input/output need arises. Then such scheduling is called *non-preemptive scheduling*. In such scheduling, once the CPU is allocated to a process, the process keeps it until its execution is terminated or it switches to the waiting state.

Dispatcher: The module of the OS that provides control of the CPU to the process which is selected by the short term scheduler is called the *dispatcher*. The dispatcher should be as fast as possible, as it is executed during every context switch. The time consumed by the dispatcher is known as *dispatch latency*.

Did you know?	Scheduling algorithms were first designed in the mid-50s.

⊙ Self-Quiz

Q3. Define preemptive scheduling?
Q4. Define dispatcher.

SCHEDULING ALGORITHMS

The computer resources are scheduled with the goal of maximizing the throughput. Some methods are preemptive whereas some are non-preemptive. Various algorithms are present which are explained below.

First Come First Served *(FCFS)* Scheduling

- Processes are executed on the first come first served basis.
- The newly created processes are placed in a ready queue.
- The scheduler selects a process from the head of the queue.
- When a new process is created it is placed at the end of the queue.
- The process that arrives first in the system is selected by the scheduler and placed into the CPU for its execution.
- The process that comes early in the system always gets the chance to get executed first.
- The size of the process does not matter during selection by the scheduler.
- This type of scheduling algorithm is also known as First In First Out (*FIFO*).
- This scheduling algorithm is easily implemented.

Example 5.1: Consider the following processes along with their burst time (in milliseconds):

Process	Burst Time (in ms)
P1	15
P2	10
P3	5

Calculate the average waiting time if the processes arrive in the following order

(i) P1, P2, P3
(ii) P2, P3, P1

Solution:

(i) Case 1: Consider a case if processes P1, P2, P3 arrive **at the same time** say 0 ms in the system in the order P1, P2, P3. The arrival of processes is represented in the Gantt chart as shown below.

Gantt chart

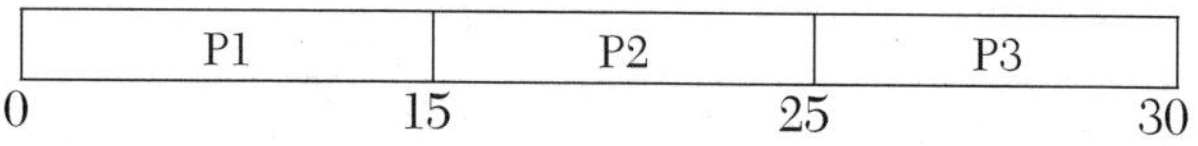

P1	P2	P3
0   15   25   30

FCFS

From the *Gantt chart,* the waiting time and the turnaround time are calculated as

Process	Waiting time	Turnaround Time
P1	0	15
P2	15	25
P3	25	30

Therefore, average waiting time = (0 + 15 + 25)/3 ms
$$= 40/3 \text{ ms}$$
$$= 15.67 \text{ ms}$$
Turnaround Time = (15 + 25 + 30)/3 ms
$$= 70/3$$
$$= 23.3 \text{ ms}$$

Case 2: Processes arrive in order P1, P2, P3, let us assume at 0 ms, 1 ms and 2 ms, respectively.

Process	Burst Time	Arrival Time
P1	15	0
P2	10	1
P3	5	2

According to FCFS (a non-preemptive algorithm), when a process starts its execution, it will not be preempted until it completes its execution. The arrival of processes is shown in the Gantt chart shown below.

Gantt chart:

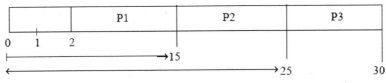

FCFS

From the Gantt chart, the waiting time and the turnaround time are calculated as:

Waiting time for process P1 = 0 ms
Waiting time for process P2 = 15 ms – 1 ms = 14 ms
Waiting time for process P3 = 25 ms-2 ms = 23 ms

Process	Waiting Time	Turnaround Time
P1	0	15
P2	14	25
P3	23	30

Therefore, the average waiting time = (0 + 14 + 23)/3 ms
$$= 37/3 \text{ ms}$$
$$= 12.3 \text{ ms}$$
Turnaround time = (15 + 25 + 30)/3 ms
$$= 70/3$$
$$= 23.3 \text{ ms}$$

(ii) If the processes P1, P2, P3 arrive **at the same time,** say 0 ms in the system in the order P2, P3, P1, then their arrival is shown in the Gantt chart below.

P2	P3	P1
0	10	15 30

From the Gantt chart, the waiting time and the turnaround time are calculated as:

Process	Burst Time	Turnaround Time
P1	15	30
P2	0	10
P3	10	15

Therefore, the average waiting time = $(15 + 0 + 10)/3$ ms
$$= 25/3 \text{ ms}$$
$$= 8.3 \text{ ms}$$
The turnaround time = $(30 + 10 + 15)/3$ ms
$$= 55/3$$
$$= 18.3 \text{ ms}$$

Drawbacks of FCFS

- The performance of this technique is lower as the average waiting time of the processes is high in this method.
- This method is non-preemptive.
- If a higher priority process arrives, then it will have to wait for its turn in the ready queue.

Shortest-Job-First (SJF) Scheduling

- Scheduler selects the process that has the smallest CPU burst time.
- The process that needs less CPU time gets priority over other processes.
- This method is of two types:

1. **Preemptive** - If a new process arrives with a CPU burst length less than remaining time of the current executing process, preempt. This scheme is known as the *Shortest-Remaining-Time-First (SRTF)*.
2. **Non-preemptive** - once the CPU is given to the process, it cannot be preempted until it finishes its CPU burst.

- SJF is the optimal method because it gives the minimum average waiting time for a given set of processes.

Example 5.2: Consider the following process along with its burst time (in milliseconds)

Process	Burst Time (in ms)
P1	15
P2	10
P3	5

Calculate the average waiting time and turnaround time using non-preemptive SJF scheduling.

Solution:

Non-preemptive SJF scheduling—according to this algorithm, the arrival of processes is shown in the following Gantt chart.

Gantt Chart

P3	P2	P1
0	5	15	30

From the Gantt chart, the waiting time and the turnaround time are calculated as

Process	Waiting Time	Turnaround Time
P1	15	30
P2	5	15
P3	0	5

Therefore, average waiting time = (15 + 5 + 0)/3 ms
$$= 20/3 \text{ ms}$$
$$= 6.6 \text{ ms}$$
Turnaround time = (30 + 15 + 5)/3 ms
$$= 50/3$$
$$= 16.6 \text{ ms}$$

Drawbacks of SJF

- This method is not a practically feasible method.
- It requires much information about the process to be scheduled.
- The CPU time required by the process should be known in advance.

Shortest-Job-Remaining-First (SJRF) Scheduling

- This algorithm is a variant of the SJF algorithm.
- This is a preemptive form of the SJF algorithm.
- Also known as *Shortest-Remaining-Time-First* (SRTF).

Example 5.3: Consider the following process along with their burst time (in milliseconds).

Process	Burst Time (in ms)	Arrival Time
P1	15	0
P2	10	1
P3	5	2

Calculate the average waiting time and turnaround time using preemptive scheduling (SJRF).

Solution:

Non-preemptive SJRF scheduling—according to this algorithm, the arrival of processes is shown in the following Gantt chart.

At T = 0 ms only one process, P1, is in the system, whose burst time is 15 ms so it begins execution of P1.

At T = 1 ms, process P2 also arrives in the ready queue along with burst time 10 which is less than the remaining time for P1 = 14, so P1 is preempted and the execution of P2 is started.

At T = 2 ms, process P3 also arrives in the ready queue along with a burst time of 5ms which is less than the remaining time for P1 = 14 and P2 = 9, so P2 is preempted and the execution of P3 is started.

Waiting time for process P1 = 0 + 16 - 1 = 15 ms
Waiting time for process P2 = 1 + (7 - 1) - 1 = 6 ms
Waiting time for process P3 = 2 - 2 = 0 ms

From the Gantt chart, the waiting time and the turnaround time are calculated as:

Process	Waiting Time	Turnaround Time
P1	15	30
P2	6	15
P3	0	5

Therefore, the average waiting time = (15 + 6 + 0)/3
$$= 21/3$$
$$= 7 \text{ ms}$$
Average turnaround time = (30 + 15 + 5)/3 ms
$$= 50/3$$
$$= 16.6 \text{ ms}$$

Priority Scheduling

Each process is assigned a priority. The priority is assigned by the operating system. The priority of a process is decided on the basis of many considerations. The considerations could include:

1. Size of the process
2. Nature of the process
3. Type of the process
4. Resource requirements of the process

This scheduling technique is of two types: preemptive and non-preemptive. In the case of preemptive scheduling, if a high priority process arrives, then the process of low priority which may be in the execution phase will have to leave the CPU. This may lead to starvation. Starvation occurs when a low priority process does not get the chance to execute because high priority processes keep arriving; then such processes are said to be in a starved state.

Example 5.4: Consider the following process along with their arrival times and priority list.

Process	Burst Time (in ms)	Priority
P1	15	3
P2	10	1
P3	5	2

Calculate the average waiting time and turnaround time using priority scheduling where the priority is high if the priority number is small, i.e., 1 > 2 > 3.

Solution:
Schedule the process according to the following Gantt chart:

P2	P3	P1

0   10   15   30

Waiting time for process P1 = 15 ms
Waiting time for process P2 = 0 ms
Waiting time for process P3 = 10 ms
From the Gantt chart, the waiting time and the turnaround time are calculated as:

Process	Waiting Time	Turnaround Time
P1	15	30
P2	0	10
P3	10	15

Therefore, the average waiting time = (15 + 0 + 10)/3
$$= 25/3$$
$$= 8.3 \text{ ms}$$
Average turnaround time = (30 + 10 + 15)/3 ms
$$= 55/3$$
$$= 18.3 \text{ ms}$$

Example 5.6: Consider the following process along with its burst time, arrival time, and priority list.

Process	Burst Time (in ms)	Priority	Arrival Time
P1	15	3	0
P2	10	1	1
P3	5	2	2

Calculate the average waiting time using priority scheduling.

Solution:
At T = 0 ms, only one process P1 is in the ready queue whose burst time is 15 ms and priority = 3.
Begin execution of P1.
At T = 1 ms, process P2 also arrives in the ready queue with the burst time 10 and priority 2.
Because the priority of P2 is higher than the priority of process P3 (priority 2 > priority 3), then the P1 process is preempted, and P2 starts its execution.

At T = 2 ms, process P3 also arrives in the ready queue with the burst time = 5 and priority 1.

Since the priority of P3 is higher than the priority of process P2 (priority 1 > priority 2), the process is preempted and P3 starts its execution.

Process P2 executes until completion, because P2 is at a higher priority among the three processes.

After completion of process P3, the P2 process starts its execution for the remaining burst time (because its priority is higher among the remaining two processes, and runs to completion. After its completion process, P1 starts its execution and runs to completion as it is the only process in ready queue.

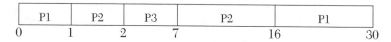

Waiting time for process P1 = 0 + 16 = 16
Waiting time for process P2 = 1 - 1 + (7 - 1) = 6
Waiting time for process P3 = 2 - 2 = 0
Average waiting time = (16 + 6 + 0)/3
= 22/3
= 7.3 ms

The drawback of the priority scheduling algorithm is that it suffers from the problem of starvation. Starvation could be removed by "aging." *Aging* means the priority of a process is increased with time. If the process is waiting for a long time, then its priority is increased.

Round Robin (RR) Scheduling

Round Robin Scheduling is based on a time sharing system and considered to be the most commonly used scheduling method. CPU time slots are made. Each process uses the CPU only in that time slot. After that, the next process in the ready queue will get the chance to execute. The incomplete process goes back to the end of the ready queue. Processes are selected from the ready queue in the FCFS manner. This technique is similar to FCFS, but it is preemptive in nature. This process continues until all other processes are terminated.

Example 5.7: Consider the following process along with their CPU burst times:

Process	Burst Time (in ms)
P1	8
P2	6
P3	5

Calculate the average waiting time and turnaround time using round robin scheduling, where the time quantum q = 2 ms.

Solution:

According to the Round Robin algorithm, the arrival of processes is shown in the following Gantt chart.

Gantt chart

P_1	P_2	P_3	P_1	P_2	P_3	P_1	P_2	P_3	P_1

0   2   4   6   8   10   12   14   16   17   19

From the Gantt chart, the waiting time and the turnaround time are calculated as:

Waiting time = turnaround time – CPU burst time
Waiting time for P1 = 19 - 8 = 11 ms
Waiting time for P2 = 16 - 6 = 10 ms
Waiting time for P3 = 17 - 5 = 12 ms

Process	Waiting Time	Turnaround Time
P1	11	19
P2	10	16
P3	12	17

Therefore, the average waiting time = (11 + 10 + 12)/3 ms
$$= 33/3 \text{ ms}$$
$$= 11 \text{ ms}$$
Turnaround time = (19 + 16 + 17)/3 ms
$$= 52/3$$
$$= 17.33 \text{ ms}$$

Drawbacks of Round Robin scheduling

■ No priority is given to processes.
■ Context switching overhead is high.

Self-Quiz

Q5. Which algorithm is used in the case of a tie between two processes?
Q6. Why is the SJF algorithm difficult to implement?
Q7. The Round Robin algorithm is a variation of which algorithm?

SOME OTHER SCHEDULING ALGORITHMS

There are some other types of scheduling techniques that can be used when processes can be divided into groups according to their types. The two main types of methods are *multilevel queue scheduling* and *multilevel feedback queue scheduling*.

The Multilevel Queue Scheduling Algorithm

In the *multilevel queue scheduling algorithm* the processes are grouped according to their types. The ready queue is divided into several small queues as shown in Figure 5.1. The group of processes is placed into each group. The upper level queue has a higher priority than the lower level. Processes cannot change their queues during their lifetimes. Every queue may use a different scheduling algorithm. The main points to remember about this technique are:

- Various queues are formed.
- Processes are divided and placed in a queue depending upon their nature.
- The queue on the topmost level is the highest priority.
- The queue at the lowest level is the lowest priority.
- Priority decreases from top to bottom.
- The different types of processes could be system processes, interactive editing processes, student process, etc.
- The scheduler selects the process from the highest level queue and places it into the CPU.
- Each queue may have its own scheduling algorithm.

Drawbacks

- This method lacks flexibility.
- Once the process is placed in a queue, it cannot change its queue.
- Processes cannot move between the queues.
- Processes in the lower queues will not be executed until processes in the higher level queues are executed and terminated.

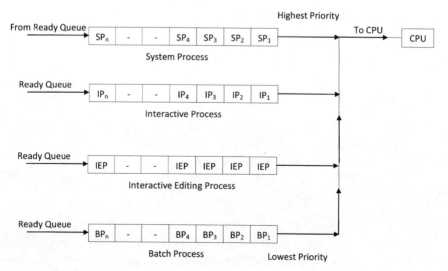

FIGURE 5.1. Multilevel Queue Scheduling

Multilevel Feedback Queue Scheduling

The drawback of the above scheduling method is that it lacks flexibility. Once the process is allocated at a level, it cannot be changed. This drawback is removed in this technique which is called *multilevel feedback queue scheduling*. In this method, the process can move between the queues. In this method, every queue is given a time quantum. Processes from the ready queue are put into the first queue. If its time requirement is less than or equal to the time quantum of that queue, then it will be executed and terminated. Otherwise the process will move to the lower level queue. Various queues and their feedback into the next queue are shown in Figure 5.2. This process continues until the process gets completed. In implementation of a multilevel feedback queue, various parameters are to be taken, which are as follows:

- Total number of queues.
- Scheduling algorithm for each queue.
- How to demote a process to a lower queue?
- How to promote a process from a lower to upper queue?

The main points to be remembered about this technique are as follows:

- Various queues are formed.
- Every queue has a specific time quantum.
- The queues are connected with each other.
- A process from the ready queue is put in first queue.
- If it finishes its execution it gets terminated; otherwise it will be moved to a lower level queue.
- Each queue may have a different scheduling algorithm.
- If the process needs a higher CPU time, it automatically moves to a lower level queue.
- Hence processes with low CPU time requirements gain an automatic priority.

◉ Self-Quiz

Q8. What are the advantages of the multilevel feedback algorithm vs. the multilevel queue algorithm?

Q9. Discuss the main features of using the multilevel queue scheduling.

MULTIPROCESSOR SCHEDULING

When many processors are present, they must be scheduled. Various processes are to be scheduled among more than one processor. Such scheduling is called *multiprocessor scheduling*.

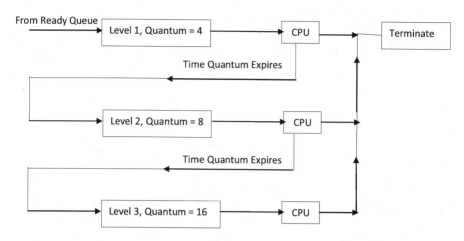

FIGURE 5.2. Multilevel Feedback Queue Scheduling

Processes can be assigned to the processors using various approaches. All of the processors are considered as pooled resources and processes are allocated to them. Two basic approaches are discussed here: (i) *static assignment,* and (ii) *dynamic assignment.*

Let us briefly discuss each:

Static assignment: In this approach, a process is assigned to one processor from its creation until its completion. A dedicated short-term queue is maintained for each processor.

Advantage: The advantage of this approach is that there is less overhead in the scheduling.

Disadvantage: The disadvantage in this approach is that one processor can be idle at a time, with an empty queue, whereas another processor may have a long queue of processes.

Dynamic assignment: In this approach, a common queue is maintained for all of the processes. They are scheduled to any available processor. Thus, in its life cycle, the process may be executed on different processors at different times.

Advantage: The advantage of this approach is that there is better processor utilization.

Disadvantage: The disadvantage of this approach is that communication between processes is difficult, because:

■ The processors could be in *master-slave* relationship or they could be in peer-to-peer relationship.
■ In a master-slave, the master processor will be responsible for the scheduling of the processes among various processors.
■ If the master processor fails, then the whole system may come to a halt.
■ In peer-to-peer, the processes are self-scheduling.

REAL-TIME SCHEDULING

A *real-time* operating system has a pre-specified maximum limit for each task that it performs to support applications with precise timing requirements. Systems that can guarantee these maximum times are called *hard real-time systems*. The systems that meet these time deadlines most of the time are called *soft real-time systems*.

Real-time systems have the following characteristics:

- Schedulers with priority-based support
- Guaranteed maximum time to service interrupts
- Ability to ensure that processes are fully loaded into memory and stay in memory
- Consistent and efficient memory allocation
- System calls that are preemptive

In *real-time scheduling*, the algorithms should be able to share a processor or any resource by processes according to some timing requirements. They should have analytical methods, also known as *feasibility tests* or *schedulability tests*, that let a system designer to analyze and "compute" the system behavior before execution time.

The algorithms for real-time scheduling should fulfill two main points:

- They should be predictable.
- While doing the scheduling, the urgency and importance of the process should be considered.

Types of Real-Time Schedulers

There are many types of schedulers for performing real-time scheduling.

- *Online/off-line scheduler*: In an *online scheduler*, the scheduling is done at execution time, whereas in an *offline scheduler* it is done before the execution time of processes.
- *Static/dynamic priority scheduler*: When priorities change at execution time, they are called *dynamic schedulers* or otherwise called *static schedulers*.
- *Preemptive or non-preemptive scheduler*: Preemptive schedulers are more efficient than *non-preemptive schedulers* because in non-preemptive scheduling there could be missed deadlines of high priority processes.

SUMMARY

Scheduling is an operation of selecting a process that is to be serviced by the CPU. A good scheduling technique is very important to decide the performance of the system. Various scheduling techniques exist that influence the performance of the computer system. The main algorithms are FCFS, SJF, Priority, and Round Robin. Every scheduling technique has its advantages and disadvantages. *The algorithm which maximizes the throughput is considered to be the best solution.*

POINTS TO REMEMBER

- Scheduling is the operation of selecting a process which is to be serviced by the CPU.
- The switching of the CPU from one process to another, whenever any need in a process arises, other than processing, is called CPU scheduling.
- The selection criterion of any algorithm depends upon CPU utilization, throughput, turnaround time, waiting time and response time.
- The main algorithms are FCFS, SJF, Priority, and Round Robin.
- When the CPU switches from one process to another before its completion, it is called preemptive scheduling.
- The CPU executes the process until it gets terminated or until any input/output need arises in it. Then this scheduling is called non-preemptive scheduling.
- In FCFS scheduling, processes are executed on the first come first served basis.
- In SJF, the process which needs less CPU time is executed first.
- In a priority algorithm, the process with highest priority is executed first.
- In the round robin algorithm each process is executed for a predefined time slot and its turn will come next after all processes in that queue get their timeslots at least once.
- In a multilevel queue algorithm, processes are divided and placed in a queue depending upon their nature.
- Each queue may have its own scheduling algorithm.
- In the multilevel feedback queue algorithm, the queues are connected with each other.
- A process from the ready queue is put in the first queue.
- If it finishes its execution, it gets terminated; otherwise it is moved to a lower level queue.
- Each queue may have a different scheduling algorithm.
- If the process needs more CPU time, it automatically moves to a lower level queue.
- Various processes are to be scheduled among more than one processor. Such scheduling is called multiprocessor scheduling.

REVIEW QUESTIONS

Q1. What do you understand by CPU scheduling? Which one is best and why?

Q2. Calculate the turnaround time and waiting time for the following processes, if these processes are using:

   i. SJF
   ii. FCFS

Q3. Discuss the multiprocessor scheduling.

Process	Arrival Time	Burst Time
P1	0	8
P2	1	4
P3	2	9
P4	3	5

Q4. Discuss the functioning of multilevel feedback queue scheduling.

Q5. Consider the following four processes with the length of CPU burst time in milliseconds.

Process	Arrival Time	Burst Time
P1	0	11
P2	2	7
P3	3	12
P4	4	8

Find which of the following algorithms gives the least average waiting time.

  i.   FCFS
  ii. RR (time slice = 5)
  iii. SRTF

Q6. Discuss the performance criteria for CPU scheduling.

Q7. Five processes A, B, C, D, E require CPU bursts 3, 5, 2, 5, and 5 respectively. Their arrival times in the system are 0, 1, 3, 9, and 12, respectively. Draw the Gantt chart and compute the average turnaround time and average waiting time of these processes for the shortest job first (SJF) and the shortest remaining time first (SJRF) scheduling algorithm.

Q8. What is the difference between preemptive and non preemptive scheduling?

Q9. Consider the following process:

Process	Arrival Time	Burst Time
P1	0.0	7
P2	2.0	4
P3	4.0	1
P4	5.0	4

Consider non-preemptive and preemptive SJF algorithms, and find out average waiting time in both cases.

Q10. Consider the set of processes given in the table and the following scheduling algorithm

  i.   Round Robin (Quantum = 1)
  ii. Round Robin (Quantum = 2)
  iii. Shortest Job Remaining First

Process Id	Arrival Time	Execution Time
A	0	4
B	2	7
C	3	3
D	3.5	3
E	4	5

If there is a tie within the processes, the tie is broken in favor of the oldest process. Draw the Gantt chart and find the average waiting time, the average response time and the turnaround time for each of the algorithms. Comment on your result and the advantages of each?

CHAPTER 6

PROCESS SYNCHRONIZATION

Purpose of the Chapter

The purpose of this chapter is to describe the inter-process communication mechanism, to introduce the critical-section problem, and to present both the software and the hardware solutions of the critical-section problem.

INTRODUCTION

Earlier, computers were allowed to execute only one task at a time. As a result, that program had full control of the system and had access to all of the system's resources. Now, however, computer systems are allowed to execute multiple programs concurrently. As a result, more security and more compartmentalization of the various programs are needed. Thus, the concept of process arises (a process is a program under execution). A *process* is a unit of work in a modern time-sharing system.

A *cooperating process* is one that can affect or can be affected by other processes that are executing in the system. Cooperating processes can either directly share a logical address space or be allowed to share data only through files or messages. The former case is achieved through the use of lightweight processes or threads. In this chapter, we discuss various mechanisms to ensure the orderly execution of cooperating processes that share a logical address space, so that data consistency is maintained.

INTER-PROCESS COMMUNICATION

In the operating system, the processes that are executing concurrently are either *independent processes* or cooperating processes. Any process is considered as independent if it cannot affect other processes executing in the system or can be affected by those processes. In other words, we can say that, a process that *does not* share data with any other process is an independent process, whereas it is defined as a cooperating process if it can affect or be affected by the other processes that are executing in the system. Therefore, cooperating processes are those processes that share data with other processes. Reasons for providing an environment that allows process cooperation follow:

- **Information sharing:** When several users may be interested in the same portion of information (for example, a shared file), then an environment could be provided that allows concurrent access to such information.
- **Computation speed up:** If any specific task is required to run faster, then it should be partitioned into subtasks. Each of these subtasks will execute in parallel with each other.

  Note: Such a speed up can be achieved only if the computers are having multiple processing elements (such as Central Processing Units or Input / Output channels).

- **Modularity:** A system can be constructed in a *modular fashion* by dividing the system functions into separate processes or threads.
- **Convenience:** Multiple tasks can be done at the same time by an individual user. For example, a user may edit, print, and compile in parallel at the same time.

A mechanism that allows cooperating processes to exchange data and information is known as an ***Inter-process communication*** *(IPC)*. The two fundamental models of inter-process communication are:

- **Shared Memory Model:** In this model, a region of memory is established that can be shared by cooperating processes. These processes can exchange information by reading and writing data to the shared region.
- **Message Passing Model:** In this model, processes can communicate with each other by exchanging messages among cooperating processes. This model is useful in exchanging small amounts of data.

Comparison of the Two Models

- The two communications models are contrasted in Figure 6.1.
- For inter-process communication, implementing message passing is easier than implementing shared memory.
- Compared to message passing, shared memory is faster.
- *Message passing systems* are implemented using system calls that involve the more time consuming task of kernel intervention. Whereas, in *shared memory*, system calls are required to establish shared memory regions. After the establishment of this shared memory, all the accesses are treated as routine memory accesses, and no support from the kernel is required.

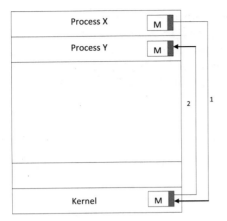 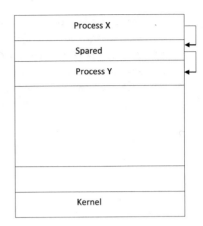

FIGURE 6.1. Communications Models, (a) Message Passing, (b) Shared Memory

Shared Memory Systems

Inter-process communication that uses shared memory requires communicating processes to establish a region of shared memory. This shared-memory region resides in the address space of the process that created the shared memory segment. Other processes that need to communicate using this shared memory segment must attach it to their address spaces. For the purpose of illustrating the concept of cooperating processes, let's consider the producer-consumer problem, which is a common example for cooperating processes.

Producer-Consumer Problem

A *producer process* creates information, and that information is consumed by a consumer process. For example, a compiler produces assembly code, which is consumed by an assembler and the assembler produces object modules that are consumed by the loader. The producer-consumer problem also provides a useful image for the client/server model. The server can be considered as a producer and a client can be considered as a consumer.

There are several solutions for the producer-consumer problem. One solution uses shared memory. For concurrent running of the producer and the consumer processes, a buffer of items is required. This buffer can be filled by the producer and get emptied by the consumer. This buffer resides in a shared memory region, and this region is shared by the producer and the consumer processes. A producer can create one item while the consumer is consuming another item. There must be synchronization between the producer and consumer, so that the consumer does not try to consume an item that has not yet been produced.

Buffers can be of two types: (i) unbounded and (ii) bounded. In the *unbounded buffer*, there is no limit on the size of the buffer. The consumer may have to wait for new items, but the producer can always create new items, whereas in the *bounded buffer*, the buffer size is fixed. In this case, if the buffer is empty, the consumer must wait, and if the buffer is full, the producer must wait.

How can the bounded buffer be used to enable processes to share memory?

The following are the variables that reside in a shared memory region, that is shared by the producer and consumer processes:

```
#define SIZE_OF_BUFFER 10
typedef struct {
} item;
item buffer [SIZE_OF_BUFFER] ;
int in = 0,-
int out = 0 ;
```

The *shared buffer* can be implemented as a circular array with two logical pointers: "in" and "out," where "in" points to the next free position in the buffer; and "out" points to the first full position in the buffer. The buffer is empty when in == out; and the buffer is full when ((in + 1) % SIZE_OF_BUFFER) == out.

Programs 6.1 and 6.2 represent the code for the producer and consumer processes, respectively. The producer process has a local variable next_Produced, which stores the new item to be produced. The consumer process has a local variable next_Consumed, which stores the item to be consumed. This scheme allows maximum of SIZE_OF_BUFFER- l items in the buffer at the same time.

PROGRAM 6.1: *The producer process*

```
item next_Produced;
while (true) {
    /* produce an item in next_Produced */
    while (((in + 1) % SIZE_OF_BUFFER) == out)
    ; /* do nothing */
    buffer[in] = next_Produced;
    in = (in + 1) % SIZE_OF_BUFFER;
}
```

PROGRAM 6.2: *The consumer process*

```
item next_Consumed;
while (true) {
    while (in == out)
    ; //do nothing
    next_Consumed = buffer[out];
    out = (out + 1) % SIZE_OF_BUFFER;
    /* consume the item in next_Consumed */
}
```

Message-Passing Systems

The *message passing system* provides a mechanism that allows the processes to communicate and to synchronize their actions with each other without sharing the same address space. It is useful in a distributed environment, where the communicating processes may reside on different computers connected by a network. For example, a *chat program* used on the World Wide Web (WWW) could be designed so that chat participants communicate with one another by exchanging messages.

A message passing facility provides two operations, i.e., send (message) and receive (message). The messages sent by a process can be either of fixed or of variable size. If only *fixed-sized messages* are sent by the process, then the system-level implementation is straightforward (which makes the task of programming more difficult). On the other hand if *variable-sized messages* are sent, then the system-level implementation is more complex (but the programming task becomes simpler).

If two processes P_1 and P_2 want to communicate with each other, then they must send messages to, and receive messages from, each other, and a *communication link* must be established between the processes. There are various methods to logically implement this communication link using send()/receive() functions by:

1. Direct or indirect communication
2. Synchronous or asynchronous communication
3. Automatic or explicit buffering

Naming

Processes that communicate with each other can either use direct communication or indirect communication.

When *direct communication* is used, then each process that wants to communicate must clearly specify the recipient or sender of the communication. Under the direct communication mechanism, the send() and receive () primitives are defined as:

- Send (P_1, message)—Send a message to process P_1.
- Receive (P_2, message)—Receive a message from process P_2.

Under this scheme, the communication link has the following properties:

- A link should be established automatically between each pair of processes that want to communicate. The processes have information only about each other's identity to communicate.
- A link should be associated with exactly two processes.
- There exists exactly one communication link between each pair of processes. In this scheme, both the sender process and the receiver process must name the other to communicate, i.e., there is symmetry in addressing. An alternative of this scheme employs *asymmetry in addressing*, i.e., only the sender names the recipient, whereas the recipient is not required to name the sender. In this scheme, the send() and receive() primitives are defined as follows:
- Send (P_1, message)—Sends a message to process P_1.
- Receive (ID, message)—Receives a message from any process, and the variable "ID" is set to the name of the process with which communication has taken place.

The disadvantage of both of these schemes (i.e., symmetric and asymmetric) is the limited modularity of the resulting process definitions. Changing the identifier of a process might require examining all other process definitions.

When indirect communication is used, the messages are sent and received through mailboxes, or *ports*. These mailboxes are considered as objects into which messages can be cited by the processes and from which messages can be removed. Each mailbox contains a unique identification. In indirect communication, a process can communicate with some other processes via a number of different mailboxes. The communication between two processes is possible only if the processes have a shared mailbox. For this scheme the send() and receive() primitives are defined as follows:

■ Send (X, message)—Send a message to mailbox X.
■ Receive (X, message)—Receive a message from mailbox X.

In the indirect communication scheme, a communication link has the following properties:

■ A link is established between a pair of processes, if both of the processes have a shared mailbox.
■ A link could be associated with more than two processes.
■ There may be a number of different links between each pair of communicating processes, where each link corresponds to one mailbox.

Let us consider that there are three processes P_1, P_2, and P_2. All of these processes share a mailbox X. Now if Process P_1 sends a message to X, while both P_2 and P_3 execute a receive() from X, then knowing which of the processes will receive the message sent by P_1 depends on which of the following methods is chosen:

■ Allows a link to be associated with no more than two processes.
■ Allows no more than one process at a time to execute a receive() operation.
■ Allows the system to select randomly which process will receive the message (i.e., either P_2 or P_3, but not both). An algorithm may be defined by the system for selecting which process will receive the message, by which the system may identify the receiver to the sender.

A mailbox can be maintained either by a process or by the operating system. If the mailbox is maintained by a process, then it is easy to differentiate between the owner and the user. Because each mailbox has a unique owner, then there is no misperception about who should receive a message sent to this mailbox. When a process that maintains a mailbox terminates, the mailbox also disappears, and if any other process sends a message to this mailbox, then that process knows the mailbox no longer exists.

On the other hand, if a mailbox is maintained by the operating system, then it is independent, and is not attached to any process. In this case, the operating system must provide a mechanism that allows a process to do the following:

■ Create a new mailbox.
■ Send and receive messages through the mailbox.
■ Delete a mailbox.

By default a process that creates a new mailbox is the owner of that mailbox, which initially is the only process that can receive messages through this mailbox. This ownership and receiving privilege may be passed to other processes through appropriate system calls. However, this establishment could result in multiple receivers for each mailbox.

Synchronization

Communication between processes takes place by calling send() and receive() primitives. For implementing each of these primitives, there are different design options.

Message passing may be either *blocking* or *non-blocking*, which is also called *synchronous* and *asynchronous*.

- **Blocking send:** In this, the sending process is blocked until the sent message is received by the receiving process or by the mailbox.
- **Non-blocking send:** In this, the sending process sends the message and resumes its operation.
- **Blocking receive:** In this, the receiver is blocked until a message is available.
- **Non-blocking receive:** In this, the receiver retrieves either a valid message or a null.

Using *blocking* and *non-blocking* schemes, different combinations of send() and receive() are possible. Then both the send() and receive() are blocking, then *rendezvous* between the sender and the receiver. When blocking send() and receive() is used, then the solution to the producer-consumer problem becomes trivial. The producer only invokes the blocking send() and waits until the message is delivered either to the receiver or to the mailbox. Similarly, when the consumer invokes receive(), it blocks until the message is available.

Buffering

Whether there is a direct communication, or indirect communication among processes, the messages are exchanged by communicating processes and these messages reside in a temporary queue. The following three methods are used to implement these queues:

- **Zero capacity:** In this method, the maximum length of the queue is zero. Therefore there are no messages waiting in the link. In this case, the sender is blocked until the receiver receives the message.
- **Bounded capacity:** In this method, the length of queue is finite, i.e., l. Therefore a maximum of l messages can reside in the queue. Whenever a new message is sent, it is placed in the queue; if the queue is empty, the sender must be blocked until the space is available in the queue.
- **Unbounded capacity:** In this method, the length of the queue is infinite. Therefore, any number of messages can wait in the queue and the sender is never blocked.

The zero capacity case is referred to as a message system with no buffering, whereas the other cases are referred to as systems with automatic buffering.

(•) Self Quiz

Q1. Discuss inter-process communication.
Q2. Explain both the models of inter-process communication.
Q3. Discuss the properties of a communication link in the indirect scheme.
Q4. State the difference between zero capacity and bounded capacity.

PROCESS SYNCHRONIZATION

We have discussed that a model is developed for a system that consists of cooperating sequential processes or threads.

All of these processes run asynchronously and might share data. This model is demonstrated with the producer-consumer problem. In the previous section, we studied how a bounded buffer could be used to enable processes to share memory. This solution allows a maximum of SIZE_OF_BUFFER-1 items in the buffer at the same time. If one needs to remove this insufficiency, then some modifications must be completed in the algorithm.

One method used to modify the algorithm is to add an integer variable counter "C," which is initialized to 0. C is incremented every time a new item is added to the buffer and is decremented when an item is removed from the buffer. Now the modified producer process will be as follows:

```
while (true)
{
    /* produce an item in next_Produced */
    while (C ==SIZE_OF_BUFFER)
    ; /* do nothing */
    buffer[in] = next_Produced;
    in = (in + 1) % SIZE_OF_BUFFER;
    C++;
}
```

The modified consumer process will be as follows:

```
while (true)
{
    while (C == 0)
    ; /* do nothing */
    next_Consumed = buffer [out];
    out = (out + 1) % SIZE_OF_BUFFER;
    C--;
    /* consume the item in next_Consumed */
}
```

Although both the producer and consumer processes routines are correct, they cannot function correctly when these processes are executed concurrently. For example, let's say the value of "C" is 5 and the producer and consumer processes execute the statements "C++" and "C--" concurrently. After the execution of these two statements, the value of "C" may be 4, 5, or 6, which is not appropriate, whereas the correct result is C = = 5, which is generated correctly if the producer and consumer processes execute separately.

The following steps have to be taken to show that the value of counter might be incorrect. The statement "C++" may be implemented in the machine language as (where R represents the register and C represents the counter):

$$R_1 = C$$
$$R_1 = R_1 + 1$$
$$C = R_1$$

where R_1 is a local CPU register. Similarly, the statement "C--" is implemented as follows:

$$R_2 = C$$
$$R_2 = R_2 - 1$$
$$C = R_2$$

Where R_2 is again a local CPU register. Although R_1 and R_2 might be the same physical register (according to the accumulator), the contents of these registers will be saved and restored by the interrupt handler.

The concurrent execution of "C++" and "C--" is equivalent to a sequential execution where the lower-level statements are interleaved in some arbitrary order. One such interleaving can be:

TO	Producer execute	$R_1 = C$	$\{R_1 = 5\}$
T1	Producer execute	$R_1 = R_1 + 1$	$\{R_1 = 6\}$
T2	Consumer execute	$R_2 = c$	$\{R_2 = 5\}$
T3	Consumer execute	$R_2 = R_2 - 1$	$\{R_2 = 4\}$
T4	Producer execute	$C = R_1$	$\{C = 6\}$
T5	Consumer execute	$C = R_2$	$\{C = 4\}$

Note that an incorrect state "C = 4" is encountered, which indicates that 4 buffers are full, whereas 5 buffers are actually full. When the order of the statements at T4 and T5 are reversed, then again an incorrect state is encountered which says "C = 6."

Such incorrect states have resulted because both the processes are allowed to manipulate the variable counter "C" concurrently. Such situations, where several processes are allowed to access and manipulate the same data concurrently, and the outcome of the execution depends on the particular order (the order in which the access takes place), is called a *race condition*. To guard against the race condition, it is required to ensure that only one process at a time can manipulate the variable counter "C." For this purpose the processes should be synchronized.

 Self Quiz

Q5. Define process synchronization?
Q6. What is race-condition?

THE CRITICAL-SECTION PROBLEM

Let us consider a system that consists of n processes $\{P_1, P_2, P_3, \ldots P_n\}$. Each process contains a segment of code, known as a *critical-section*, where the process might be exchanging common variables, updating a table, writing a file, etc.

One important feature of the system is that when one process is executing in its critical section, no other process is allowed to execute in its critical section. Therefore, two processes are not allowed to execute their critical sections at the same time. The *critical-section problem* designs a protocol that can be used by the processes for the purpose of cooperation. Each process must request permission to enter into its critical section. The section of code that implements this request is referred as the *entry section*. The critical section is followed by an exit section. The remaining code is known as the *remainder section*. The general structure of a process P_i, is shown in Program 6.3.

PROGRAM 6.3: General structure of a process P_i

```
do{
    entry section
    critical section
    exit section
    remainder section
} while (TRUE);
```

The three requirements that should be satisfied by a solution to the critical-section problem are:

1. **Mutual exclusion:** If any process P is executing in its critical section, then no other process is allowed to execute in their critical sections.
2. **Progress:** If no process executes the critical-section and if some processes request to enter their critical sections, then the processes that are not executing in their remainder sections can participate in the decision on which process will enter its critical section next (this selection cannot be postponed indefinitely).
3. **Bounded waiting:** A limit exists on the number of times that other processes are allowed to enter their critical sections, after a process has made a request to enter its critical section and before that request is granted.

It is assumed that each process is executing at a non-zero speed, and no assumption can be made concerning the relative speed of the n processes.

At any instance, many kernel-mode processes may be active in the operating system. Thus, the code that implements an operating system (i.e., kernel code) results in several possible race conditions. For example, consider a kernel data structure that maintains a list of all open files in the system. This list must be changed whenever a new file is opened or closed. If any two processes want to open files simultaneously, then the separate updates to this list could result in a race condition. Some other kernel data structures that result in possible race conditions are structures for maintaining memory allocation, structures for maintaining process lists, and structures for interrupt handling. Only the kernel developers can ensure that the operating system is free from such race conditions.

To handle the critical sections in operating systems, the following two approaches are used:

1. ***Preemptive kernels***: This allows a process to be preempted while it is running in kernel mode.
2. ***Non-preemptive kernel***: It does not allow a process to be preempted while it is running in kernel mode. In this approach a kernel-mode process will run until it exits kernel mode, blocks, or voluntarily yields control of the CPU.

Therefore, a non-preemptive kernel is free from race conditions on kernel data structures, because at any time only one process is active in the kernel. The same is not possible in the case of preemptive kernels, so they must be designed carefully (to ensure that shared kernel data are free from race conditions).

Although a non-preemptive kernel is free from race conditions, still a preemptive kernel is used. Because a preemptive kernel is more suitable for real-time programming, it allows a real-time process to preempt a process currently running in the kernel. A preemptive kernel may also be more responsive, since it is not possible that a kernel-mode process will run for an arbitrarily long period before surrendering the processor to waiting processes. Obviously, this effect can be minimized by designing a kernel code that does not behave in this way.

(◉) Self Quiz

Q7. Define critical-section?

Q8. How can the critical section problem be solved?

PETERSON'S SOLUTION

Peterson's Solution is a software-based solution to handle the critical-section problem. It is uncertain that this solution will work correctly on the modern computer architectures that perform basic machine language instructions, such as load and store.

However, we use this solution because it provides a good algorithmic description for solving the critical-section problem. This solution also clarifies the complexities that are involved in designing software that addresses the requirements of mutual exclusion, progress, and bounded waiting requirements. Peterson's solution is designed for two processes that alternate the execution between their critical sections and the remainder sections.

Two Process Solution

The two data items used by Peterson's solutions that have to be shared between the two processes are:

```
int t;
boolean FLAG[2];
```

Here the variable t indicates whose turn it is to enter its critical section. For example if t = i, then process P_i has permission to execute in its critical section. And the FLAG array is used to indicate that whether a process is ready to enter its critical section. For example, if FLAG[i] = true, it means that a process P_i is ready to enter into its critical section. The algorithm is depicted in Program 6.4.

For the purpose of entering into its critical section, process P_i first sets FLAG[i] = true and then sets t = j, thus stating that if the other process P_j wishes to enter into its critical section, it can do so. If both the processes try to enter into their critical sections at the same time, then "t" will be set to both i and j at the same time. But only one of these assignments will last (the other assignment will also occur, but it will be overwritten immediately). Which of the two processes has permission to enter into its critical section first is determined by the value of "t."

PROGRAM 6.4: The structure of process P_i in Peterson's solution.

```
do {
    FLAG[i] = true;
    t = j;
    while (FLAG[j] && t = j );
    critical section
    FLAG[i] = false;
    remainder section
} while (true);
```

To prove that the given solution is correct, the following criteria have to be fulfilled:

1. Mutual exclusion should be preserved.
2. The progress requirement should be satisfied.
3. The bounded-waiting requirement should be met.

Property 1: According to the solution it is clear that each P_i will enter into its critical section only if either FLAG[j] = false or t = i. Also if both the processes executing in their critical sections at the same time, then FLAG [i] = FLAG [j] = true. According to these two observations the processes P_i and P_j could not have successfully executed their while statements at the same time, because the value of 't' can be either i or j but it cannot be both. Hence, one of the processes (assume Pj) had successfully executed the while statement, whereas process P_i, had to execute one additional statement (i.e., "t = j"). At that time, FLAG[j] = true, and t = j, and this condition will persist as long as the process P_j is in its critical section. In this way mutual exclusion is preserved.

Properties 2 and 3: Any process can be prevented to enter into its critical-section only if the process is stuck in the while loop with the condition FLAG [j] = true and t = j. If the process P_j is not ready to enter into its critical-section, then FLAG [j] = false, and process P_i will be allowed to enter into its critical-section. If P_j has set FLAG[j] = true and P_j is also executing in its while statement, then the value of 't' will be either i or j. If t = i, then P_i will enter into its critical-section otherwise P_j will enter into its critical-section. Once P_j exits from its critical section, it will reset FLAG[j] = false, and allows process P_i, to enter its critical section. If P_j resets FLAG[j] = true, then t = i should also be reset. Therefore, since P_i, does not change the value of the variable 't' during the execution of its while statement, hence, P_i will enter into its critical section (i.e., progress) after at most one entry by P_j (i.e., bounded waiting).

PROGRAM 6.5: Solution to the critical-section problem using locks.

```
do {
    acquire lock
    critical section
    release lock
    remainder section
} while (TRUE);
```

⊙ Self-Quiz

Q9. Explain Peterson's solution for critical-section problem.
Q10. Write any two features of the Peterson solution.
Q11. Write a program to find the solution of the critical-section problem using locks.

DEKKER'S ALGORITHM

This algorithm is the first correct solution proposed for the case of two-thread (or two-process). Dekker developed this algorithm for different environment. Dijkstra uses this algorithm and applies it to the critical section problem.

In the context of the solution to critical-section problem, Dekker adds the idea of a chosen thread and allows access to either thread when the request is accepted. Whenever there is a conflict, one thread is chosen, and the priority reverses after successful execution of the critical-section. The variables that will be shared are:

```
int FLAG[2];
FLAG[0] = FLAG[1] = 0;
int t = 0;
```

Program 6.6 shows the structure of thread.

PROGRAM 6.6: The structure of Thread Ti in Dekker's solution.

```
do {
    FLAG[i] = 1;
    while (FLAG[j] )
    if (t == j)
    {
        FLAG[i] = 0;
        while (t == j);
        FLAG[i] = 1;
    }
    critical section
        t = j;
        FLAG[i] = 0;
    remainder section
} while (1);
```

⊙ Self-Quiz

Q12. Explain Dekker's algorithm.
Q13. Write the set of statements to show the sharing of variables applying in Dekker's solution.

SYNCHRONIZATION HARDWARE

The software-based solution to the critical-section problem was described in previous sections. To improve the efficiency of system and to make programming task easier, hardware features can be used. This section provides some simple hardware instructions and show that how these instructions can be used effectively to solve the critical-section problem.

The solution to a critical-section problem could be simple in a uniprocessor environment if the occurrence of interrupts could be prevented while a shared variable was being modified. But in the case of multiprocessor environment, this solution is not feasible. To prevent the occurrence of interrupts on a multiprocessor is time consuming, because the message has to pass to all the processors. This message passing delays the entry into the critical section of processes, and the efficiency of system decreases. If the clock is updated by the interrupt, then system's clock should also be considered.

Therefore special hardware instructions are provided by modern computer systems that allows either to *test and set* the content of a word or to *swap* the contents of two words atomically. These special hardware instructions are used to solve the critical-section problem. This section describes the main concepts behind these types of instructions.

Program 6.7 represents the TestAndSet() instruction. The main feature of this instruction is that it executes atomically. If there are two TestAndSet() instructions (each on different CPU), and both are executed simultaneously, then they will execute sequentially in some arbitrary order. If the machine is supporting the TestAndSet() instruction, then the mutual exclusion can be implemented by declaring a boolean variable lock, which will be initialized to false. Program 6.8 shows the structure of process P.

PROGRAM 6.7: The definition of the TestAndSet() instruction.

```
boolean TestAndSet (boolean *target)
{
    boolean rv = *target;
    *target = true;
    return rv;
}
```

PROGRAM 6.8: Mutual-exclusion implementation with TestAndSet().

```
do {
    while (TestAndSetLock (&lock) )
    ; // do nothing
    // critical section
    lock = FALSE;
    // remainder section
} while (TRUE);
```

Whereas the Swap() instruction operates on the contents of two words and is defined as shown in Program 6.9. Similar to the TestAndSet() instruction, Swap() instruction also executes atomically.

PROGRAM 6.9: The definition of the Swap() instruction.

```
void Swap (boolean *a, boolean *b)
{
    boolean temp = *a;
    *a = *b;
    *b = temp;
}
```

If the machine is supporting Swap() instruction, then the mutual exclusion can be provided as follows:

A global boolean variable lock is declared and it is initialized to false. Each process is having a local boolean variable key. Program 6.10 shows the structure of process P_i. Although the mutual-exclusion requirements are satisfied by these algorithms satisfy, but they do not satisfy the requirement of bounded-waiting.

PROGRAM 6.10: Mutual-exclusion implementation with the Swap() instruction.

```
do
{
    key = true;
    while (key == true)
    Swap (&lock, &key);
    // critical section
    lock = false;
    // remainder section
}while (true);
```

Program 6.11 specifies alternative algorithm that uses the TestAndSet() instruction and satisfies all the requirements of critical-section. For this algorithm the common data structure is declared as follows:

boolean waiting[n];
boolean lock;

Both the data structures are initialized to false. To prove that the requirement of mutual exclusion is satisfied, it should be noted that process P_i can enter into its critical section if either waiting[i] = false or key = false. The value of key will be set to false only if the TestAndSet() instruction is executed. The first process that execute the TestAndSet() instruction will find key = false and all the other processes must wait. The variable waiting[i] will be set to false only if another process leaves its critical section. At a time only one waiting [i] = false, which fulfills the requirement of mutual-exclusion.

PROGRAM 6.11: Bounded-waiting mutual exclusion with TestAndSet().

```
do {
    waiting [i] = true;
    key = true;
    while (waiting[i] && key)
    key = TestAndSet(&lock);
    waiting [i] = false;
    // critical section
    j = (i + 1) % n;
    while ((j != i) && !waiting[j])
    j = (j + 1) % n;
    if (j==i)
    lock = false;
    else
    waiting[j] = false;
    // remainder section
}while (true);
```

To prove that the requirement of progress is satisfied, a process that is exiting its critical section will either set lock = false or set waiting[j] = false.

Both the argument allows a process, which is waiting to enter its critical section, to proceed.

To prove that the requirement of bounded-waiting is satisfied, it should be noted that, when a process leaves its critical section, it scans the array *"waiting"* in the cyclic ordering. It labels the first process in this ordering as the next process to enter the critical section.

(•) Self-Quiz

Q14. Explain Test And Set hardware instruction.
Q15. How to solve critical-section problem?

SEMAPHORES

To solve the critical section problem, various hardware-based solutions were provided in the previous section. The use of these solutions is complicated for application programmers. This difficulty can be resolved by using a synchronization tool, known as semaphore.

A semaphore 'S' is an integer variable that is accessed only through two standard atomic operations, i.e., wait() and signal(). The wait() operation originally defined as P (from the Dutch words problem, "to test") and signal() was originally defined as V (from Verhogen, "to increment"). The wait() operation is defined as follows:

```
Wait (S) {
    while S <= 0
    ; // no operation
    S—;
}
```

The signal() operation is defined as follows:

```
signal(S) {
    S + +;
}
```

Modifications that have been made to the integer value of semaphore in the wait() and signal() operations are executed indivisibly, which means that, when one process is modifying the semaphore value, then no other process can modify the value of same semaphore simultaneously. Also, in the case of wait(S), the testing of the integer value of semaphore S (i.e., S<=0), and its possible modification (i.e. S—), must also be executed without any interruption.

Usage of Semaphores

Semaphores can be of two types, i.e., counting semaphores and binary semaphores. Counting semaphores' value can range over an unrestricted domain, whereas binary semaphores' value can range only between 0 and 1.

Binary semaphores are known as mutex locks on some systems. Mutex locks are considered as locks that provide mutual exclusion.

Binary semaphores can be used to deal with the problem of critical-section for multiple processes. Let there are n processes, all the n processes share a semaphore, mutex. In this semaphore, the mutex is initialized to 1. Program 6.12 shows that each process P_i is organized.

PROGRAM 6.12: Mutual-exclusion implementation with semaphores.

```
do {
    waiting(mutex);
    // critical section
    signal(mutex);
    // remainder section
}while (true);
```

Counting semaphores are used to control the access to a given resource that consists of a finite number of instances. The value of the semaphore is initialized to the number of resources available. The process that wants to use a resource performs a wait() operation on the semaphore (decrements the counter), whereas whenever a process releases a resource, it performs a signal() operation (increments the counter). When the value of the counter for the semaphore becomes 0, it means that all the resources are being used. Which implies that those processes, who wants to use a resource will be blocked until the value of the counter becomes greater than 0.

Semaphores can also be used to solve various synchronization problems. For example, consider two processes that are running concurrently, i.e., P_1 with a statement S_1 and P_2 with a statement S_2. Suppose it is required that S_2 will be executed only after completion of S_1. This scheme can be implemented by considering that the processes P_1 and P_2 will share a common semaphore, synch, which is initialized to 0, and by inserting the following statements:

```
S1;
signal(synch);
in the process P1, and the statements
wait(synch);
S2;
```

in the process P_2. Since synch is initialized to 0, therefore P_2 will execute S_2 only after P_1 has invoked signal (synch), which is after the execution of statement S_1.

Implementation

The major disadvantage of semaphore is its requirement of busy waiting. This means that when a process (let P) is in its critical section, and at the same time if any other process (let Q) is trying to enter its critical section, then the process Q must go to continuous loop in the entry code. In multiprogramming systems, this continuous looping is a major problem, because in such a system single CPU is shared among many processes. This busy waiting wastes the CPU cycles that can used productively by some other process. Such a semaphore is known as *spinlock* (process "spins" while it is waiting for the lock). The advantage of spinlocks is that it does not require context switching when a process must wait on a lock, and a context switching may take substantial time. Therefore, spinlocks are useful whenever locks are expected to be held for short times.

The requirement of busy waiting can be recovered by modifying the definition of the wait() and signal() semaphore operations. We know that when a process P executes the wait() operation and finds that the value of semaphore is negative, then the process P must wait. On the other hand, rather than engaging the process P in busy waiting, it can block itself. This block operation will place the process P into a waiting queue (which is associated with the semaphore), and will change the state of the process to the waiting state. Afterwards the control is transferred to the CPU scheduler, which selects another process to be executed.

If a process (let P) is blocked (i.e., P is in waiting queue/waiting on a semaphore S), then it should be restarted when any other process (let Q) executes a signal () operation. The process P is restarted by a wakeup () operation, which changes the state of the process P from the waiting state to the ready state. Then the process is placed in the ready queue.

To implement semaphores under this definition, it will be defined as:

```
typedef struct {
    int val;
    struct process *list;
} semaphore;
```

Each semaphore contains an integer value, i.e., *val* and a list of processes, i.e., *list*. Whenever a process waits on a semaphore, it is added to *list*. A signal() operation removes one process from the *list* arises that process. The modified wait() semaphore operation can be defined as:

```
Wait (semaphore *S) {
    S -> val—;
    if (S -> val< 0) {
        add this process to S ->list;
        block ();
    }
}
```

And the modified signal() semaphore operation can be defined as:

```
signal(semaphore *S) {
    S -> val++;
    if (S -> val<= 0) {
        remove a process P from S ->list;
        wakeup(P);
    }
}
```

The block() operation suspends that process which invokes it and the wakeup (P) operation resumes the execution of a process P, which was blocked. Both of these operations are provided by operating system as basic system calls.

The *list* can be implemented easily by a link field in each PCB (Process Control Block). Each semaphore contains an integer value (*val*) and a pointer to a list of PCBs (*list*). To add and remove processes from the *list*, FIFO (First In First Out) queue can be used, where the semaphore contains both the head and the tail pointers to the queue. Generally, the *list* can use any queuing strategy (because correct usage of semaphore does not depend on a particular queuing strategy).

Atomically execution is the serious facet of semaphores. A critical-section problem says that the two processes cannot execute the wait() and the signal() operations at the same time on the same semaphore. The critical-section problem in a single-processor environment can be solved simply by inhibiting interrupts during the time the wait() and the signal() operations are executing. This scheme is useful for single processor environment because once the interrupts are inhibited, then the instructions from different processes cannot be interleaved. According to this scheme only the currently running process executes until the interrupts are re-enabled and the scheduler can regain its control.

In the case of multiprocessor environment, the interrupts must be disabled on every processor, because it is possible that the instructions from different processes may be interleaved. It could be a difficult task to disable interrupts on every processor. Therefore, SMP systems must provide alternative locking techniques, i.e., spinlocks, to ensure that the wait() and the signal() semaphore operations are performed atomically.

Deadlocks and Starvation

Implementing a semaphore with a waiting queue may result in a situation where two or more processes are waiting indefinitely for an event to occur (that event can be caused by anyone of the waiting processes). The situation when such a state is reached is called deadlock and the processes that are waiting indefinitely are said to be **deadlocked.**

To demonstrate this, let us consider a system that consists of two processes, P_1 and P_2, each accessing two semaphores, S_1 and S_2, set to the value 1:

P_1	P_2
wait(S_1);	wait(S_2);
wait(S_2);	wait(S_1);
.	.
.	.
.	.
signal(S_1);	signal(S_2);
signal(S_2);	signal(S_1);

Suppose that P_1 executes wait (S_1) and then P_2 executes wait (S_2). When P_1 executes wait(S_2), it must wait until P_2 executes signal(S_2). Similarly, when P_2 executes wait(S_1), it must wait until P_1 executes signal(S_1). Since these signal() operations cannot be executed, P_1 and P_2 are deadlocked.

Therefore we can say that a set of processes is in a deadlock state when every process in the queue is waiting for an event (such as *resource acquisition and release*) to be occurred, that can be caused by any another process in the set.

Another problem that is related to deadlocks is **indefinite blocking, or starvation.** It is a situation in which processes waits indefinitely within the semaphore. Indefinite blocking can occur if we use LIFO (Last In First Out) ordering to add and to remove processes from the list associated with a semaphore.

(◉) **Self-Quiz**

Q16. What is semaphore? How it can be used?
Q17. Discuss the types of semaphores.

CLASSIC PROBLEMS OF SYNCHRONIZATION

In this section, some of the synchronization problems are presented as examples of a large class of concurrency-control problems. Every new synchronization scheme is tested using these problems. Here semaphores for synchronization are used as a solution to the problem.

The Bounded-Buffer Problem

The *bounded-buffer problem* is also known as producer-consumer problem and is commonly used to explain the influence of synchronization primitives. This section presents a general structure of this scheme.

Assume that the pool is consisting of *n* buffers, and each buffer is capable of holding one item. To access the buffer pool, the mutex semaphore provides mutual exclusion. This mutex semaphore is initialized to the value 1. To count the number of empty and full buffers, empty and full semaphores are used. The empty semaphore is initialized to the value *n* and the full semaphore is initialized to the value 0.

The code for the producer process and the consumer process is shown in Program 6.13 and Program 6.14 respectively. According to the code, the producer is producing full buffers for the consumer and the consumer is producing empty buffers for the producer.

PROGRAM 6.13: The structure of the producer process.

```
do {
    // produce an item in next_p
    wait(empty);
    wait(mutex);
```

```
    // add next_p to buffer
    signal(mutex);
    signal(full);
} while (TRUE);
```

PROGRAM 6.14: The structure of the consumer process.
```
do {
    wait(full);
    wait(mutex);
    // remove an item from buffer to next_c
    signal(mutex);
    signal(empty);
    // consume the item in next-_c
} while (TRUE);
```

The Dining-Philosophers Problem

Let us consider that there are five philosophers who are spending their lives in thinking and eating. These philosophers share a common circular table that is surrounded by five chairs. Each chair belongs to one philosopher. In the center of the table there is a bowl full of rice, and there are five single chopsticks on the table (See Figure 6.2). When a philosopher is in thinking mode, he/she does not interact with his/her colleagues. When, a philosopher gets hungry, he/she tries to pick up the two chopsticks that are closest to him/her (the chopsticks that are between that philosopher and the left and right neighbors of that philosopher). At a time a philosopher may pick up only one chopstick (he/she cannot pick up a chopstick that is in the hand of a neighbor). When a hungry philosopher gets both the chopsticks at the same time, he/she eats without releasing the chopsticks. When the philosopher has finished eating, he/she puts down both the chopsticks and starts thinking again.

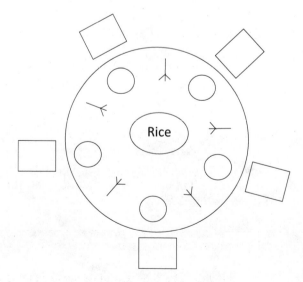

FIGURE 6.2. The Situation of the Dining Philosophers

This problem is considered a classic synchronization problem and is a simple representation of the requirement to allocate several resources among several processes in a deadlock-free and starvation-free routine.

One of the simplest solutions for this problem is to represent each chopstick with a semaphore. A philosopher will try to grasp a chopstick by executing a wait() operation on that semaphore and the philosopher will releases his/her chopsticks by executing the signal() operation on the suitable semaphores. Therefore, the shared data are:

semaphore chop_stick[5];

where all the elements of chop_stick are initialized to 1. The structure of philosopher i is shown in Program 6.15.

PROGRAM 6.15: The structure of philosopher i.

```
do {
    wait (chop_stick [i] );
    wait(chop_stick [ (i + 1) % 5] ) ;
    // eat
    signal(chop_stick [i]);
    signal(chop_stick [(i + 1) % 5]);
    / / think
} while (TRUE);
```

Though this solution does not guarantee that no two neighbors are eating simultaneously, it however must be rejected because it could create a deadlock. Assume that all the five philosophers get hungry at the same time and each philosopher grabs his/her left chopstick. Now all the elements of chopstick will be equal to 0. Hence, when any philosopher tries to grab his/her right chopstick, he/she will be delayed forever.

Some possible solutions to the deadlock problem are:

- At most four philosophers are allowed to be sitting simultaneously at the table.
- A philosopher is allowed to pick up his/her chopsticks only if both chopsticks are available (the chopsticks should be picked up in a critical-section).
- An asymmetric solution can be used, i.e., an odd philosopher will pick up his/ her left chopstick first and then pick up his/her right chopstick, whereas an even philosopher will pick up his/her right chopstick first and then pick up his/ her left chopstick.

Note: It is not necessary that a deadlock-free solution will eliminate the possibility of starvation.

The Sleeping Barber Problem

In the sleeping barber problem there is one barber chair and multiple waiting chairs.

Only one person can move at a time. Customers enter through/exit through the out door. When there is no one is in the shop, barber sleeps in the barber chair.

When a customer arrives:

- If no one else is inside the shop, they wake up the barber and waits to sit in the barber chair and get their hair cut.
- If the barber is busy, the customer waits their turn in a waiting chair. Once a haircut of the customer is finished:
- Barber opens the exit door and waits for customer to leave (close it behind them).
- Checks that is there any customers waiting? If there is none, he goes to sleep in the barber chair.
- Only one customer is waiting (and the customer is sleeping), then the barber will:

  - Wake up the customer.
  - Wait for the customer to sit down in the barber chair.

This is a very complex model and requires lots of wait() and signal() operations of semaphores.

(◉) Self-Quiz

Q18. Explain the bounded buffer problem of synchronization.
Q19. Explain the dining philosopher problem of synchronization.
Q20. Explain the sleeping barber problem of synchronization.

SUMMARY

In operating system, the executing processes may be either independent processes or cooperating processes. Cooperating processes require an inter-process communication mechanism to communicate with each other. The communication is achieved by two schemes, i.e., shared memory and message passing.

Mutual exclusion is required for a given collection of cooperating sequential processes that are sharing data. This can be solved by ensuring that at a time only one process or thread undergoes its critical section. Various algorithms are available to solve the critical-section problem. The major disadvantage of these solutions is that they all require busy waiting. To solve this busy waiting problem semaphores is used. Semaphores can be used to solve various synchronization problems and can be implemented efficiently, especially if hardware support for atomic operations is available. The synchronization problems (such as the bounded-buffer problem, the readers-writers problem, and the dining-philosophers problem) problems are used to test almost every newly proposed synchronization scheme.

POINTS TO REMEMBER

- A mechanism that allows cooperating processes to exchange data and information is known as inter-process communication (IPC).

- The two fundamental models of inter-process communication are: shared memory model and message passing model.
- Message passing may be either blocking or non-blocking.
- The situations, where several processes are allowed to access and manipulate the same data concurrently and the outcome of the execution depends on the order in which the access takes place is called a race condition.
- Critical-section is a segment of code where the process may be exchanging common variables, updating a table, writing a file, etc.
- To solve the critical section problem, the three main requirements are: mutual exclusion, progress and bounded wait.
- In OS, to handle critical-section, the two approaches are: preemptive and non-preemptive kernels.
- Semaphore is a synchronization tool.
- Two types of semaphores are available, i.e., counting and binary semaphore.

REVIEW QUESTIONS

Q1. Provide the solution of critical section problem.

Q2. Explain the binary semaphore with an example.

Q3. Write a brief note on inter-process communication.

Q4. How is the concurrency problem solved with producer-consumer problem?

Q5. Explain how readers-writers problem can be solved by using semaphore.

Q6. What is the critical section? How can we obtain a solution to the critical section problem?

Q7. Write an algorithm to explain producer/consumer using semaphores.

Q8. Discuss one classical problem related to the process synchronization.

Q9. What do you understand by critical section? What are the requirements of a solution to the critical section problem? Discuss mutual exclusion implementation with the help of Test-and-Set machine instructions.

Q10. State the finite buffer Producer-Consumer problem. Give solution of the problem using semaphore.

Q11. What do you mean by concurrent process? Discuss the inter-process communication in detail.

Q12. Write and explain the Peterson solution to the critical-section problem.

Q13. Discuss the message passing system. Explain how message passing can be used to solve Producer-Consumer problem with infinite buffer.

Q14. Define semaphore with suitable example.

Q15. Define message-passing and shared-memory inter-process communication.

Q16. List the essential requirements of critical-section problem.

Q17. State and describe the producer-consumer problem with its suitable solution.

CHAPTER 7

DEADLOCKS

Purpose of the Chapter

The purpose of this chapter is to define deadlock and describe how and why deadlock occurs. Deadlock is an unfavorable state in the system which needs to be taken care of. This chapter further explains how a deadlock can be handled in a system. Three different ways to handle and remove deadlock from the system are explained in detail.

INTRODUCTION

On a one-way road, suppose two cars come from opposite directions. Neither of them are ready to move back. If the road is so narrow that they cannot pass on the sides, then both cars will be stuck on the road until one of them moves back. This situation is called deadlock. Both cars will not be able to move further in their desired direction, and hence they are deadlocked. This situation of deadlock of the two cars is shown in Figure 7.1.

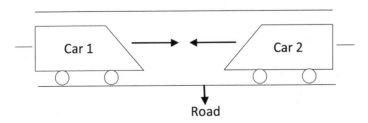

FIGURE 7.1. Two cars in deadlock state

Such a situation may also arise in the computer system. Suppose there is a set of processes which are holding a resource each and are also waiting for a resource which is held in another process. Then such a situation will create deadlock for the processes. Processes are said to deadlock in this situation.

A situation when two or more processes get blocked and cannot proceed further with their processing because of interdependency of resources is called deadlock. This situation is shown in Figure 7.2.

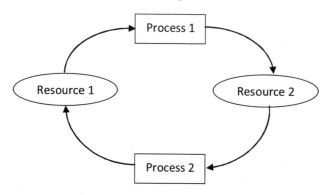

FIGURE 7.2. Two processes in deadlock state

In other words, a deadlock occurs when two or more processes attempt to access a resource that the other process has locked and hence cannot be shared. As a result, each process waits for a resource that the other has locked, and neither process can be finished.

For example, two resources, a printer and a compact disk (CD) writer drive, are attached to a computer. Two processes, P and Q, exist in a ready queue. Both processes P and Q want to read a document from the CD presently placed in the CD writer drive, make necessary changes to the document, write it back to the CD, and get a modified document printed. Both processes require both resources, the CD drive and printer, to finish their task.

The process P acquires the CD writer drive and locks it in non-sharable mode. The process Q acquires the printer and locks it in non-sharable mode. The process P waits to obtain the printer, and process Q waits to obtain the CD drive. Neither of the processes will be able to continue processing. This will lead to a hang situation, which is called deadlock.

DEADLOCK CHARACTERIZATION

In a computer system, deadlock arises when four conditions occur simultaneously. When all these conditions occur at the same time, then only a deadlock may be created among processes. The conditions are as follows:

■ **Mutual Exclusion:** Mutual Exclusion means only one at a time. This means at a particular time stamp only one process can use a resource. Multiple processes are not allowed to use the same resource at the same time. If the need occurs to access the resource simultaneously, then a queue is formed and the processes get access to the resource according to their turn.

For example, a process P locks the resource R in exclusive mode. This means no other process can acquire the resource R. In other words, the concept of mutual exclusion means that a resource can be used by only one process at a time.

■ **Hold and Wait:** When a process is holding some resources and waiting for other resources to finish execution, then such a situation is called Hold and Wait. This means a process is allowed to request new resources even if it is holding some. In other words the processes involved in a deadlock acquire a minimum of one resource in exclusive mode and wait for another resource which has been acquired by another process in exclusive mode (non-sharable mode).

■ **No preemption:** If a process is holding some resource, it cannot be taken from it until the process is finished with it. The process will release the acquired resource after completing its task. This situation is called no preemption. This means resources cannot be taken from a process forcibly. In other words processes do not release the resources allotted to them voluntarily without utilizing the resources. For example, process P1 holds the resource R1 exclusively and waits for the resource R2, which is locked by process P2 in exclusive mode. The concept of no preemption means that the operating system cannot revoke the resource R1 from process P1 forcefully.

■ **Circular wait:** Each process is waiting to obtain a resource which is held by another process in such a way so that they form a cycle. A closed chain or a circular relationship exists among the processes involved in a deadlock. For example, there are three processes, P1, P2, and P3. The process P1 is waiting for a resource that has been exclusively acquired by the process P2, process P2 is waiting for a resource which is held by P3, and in turn P3 is waiting for a resource being held by P1. This situation will result in circular wait.

(◉) Self-Quiz

Q1. Define deadlock.
Q2. What are the necessary conditions which should be present for deadlock to occur?
Q3. Why is the occurrence of deadlocks unfavorable in the system?

RESOURCE ALLOCATION GRAPH

A graph is a collection of points and lines connecting them. The points of a graph are known as nodes. The lines connecting the nodes are called edges. A resource allocation graph is a directed graph that consists of processes and resources as the two types of nodes, and indicates (a) the processes requesting resources and (b) the resources held by particular processes.

A resource allocation graph is also known as a RAG. It consists of a set of nodes P={P1, P2, ... Pn} and R={R1, R2, ... Rm} where P represents a set of processes and R represents a set of various resources present in the system. It also consists of two sets of directed edges:

- **Request Edge:** This edge extends from a process to the resource being requested by the process. It is represented as Pi → Rj.
- **Assignment Edge:** This edge extends from resource to process. It indicates which resource is allocated to which process. It is represented as Rj → Pi.

Purpose of a Resource Allocation Graph

The presence of deadlock can be indicated with the help of a RAG. In other words, a resource allocation graph is an analytical tool which is used to detect the presence of deadlock in a system. How a RAG determines the presence of deadlock in a system is described as follows:

- If the graph contains no cycles, then there is no deadlock in the system.
- If the graph contains a cycle, then there may or may not be a deadlock depending on the following conditions.
- If there is only one instance per resource type, then deadlock is present.
- If several instances per resource type are present, then there is the possibility of deadlock.

Examples of Some Resource Allocation Graphs

Figure 7.3 illustrates a RAG showing a deadlock situation, and Figure 7.4 illustrates a cycle in a RAG with no deadlock.

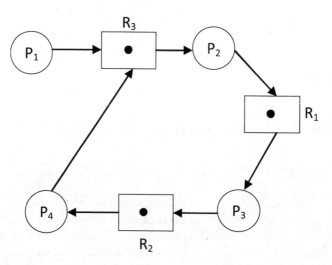

FIGURE 7.3. A Resource Allocation Graph showing a deadlock situation

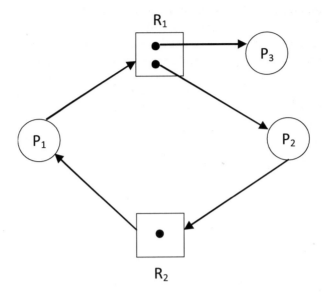

FIGURE 7.4. Cycle in a Resource Allocation Graph with no deadlock

Self-Quiz

Q4. What do you mean by a resource allocation graph?
Q5. Name some components of a RAG.
Q6. Name the edges which are present in a RAG.
Q7. What is the purpose of a resource allocation graph?

METHODS FOR HANDLING DEADLOCKS

Occurrence of deadlock in a system is not a favorable situation. There are various ways to handle deadlock. Different operating systems follow different methods to deal with the problem of deadlock, but the main methods are as follows:

■ Deadlock prevention
■ Deadlock avoidance
■ Deadlock detection

Let us discuss each one in brief.

DEADLOCK PREVENTION

As explained in Section 7.1, there are four conditions which should hold simultaneously for deadlock to occur in the system. If any of these conditions can be made false, then deadlock can be prevented. These conditions can be prevented in the following ways:

Mutual Exclusion: When resources are sharable it is not necessary for this condition to be present. For example, read-only files can be used by processes at the same time, so there is no need for mutual exclusion.

But in the case of non-sharable resources, it is necessary for the mutual exclusion conditions to be present. For example, this includes resources like printers or disk drives. Such resources can handle requests from one process at a time. Hence mutual exclusion is required.

Hold and Wait: The system should make sure that whenever a process requests a new resource, there should not be any resources in its hold. This means that a process should never hold a resource and request new ones at the same time. This can be achieved by providing all the resources in advance. All the resources required by the process during its lifetime are allocated to it before starting its execution. In this way, the condition of Hold and Wait can be prevented and hence deadlock can be prevented.

But this is not favorable, as it will lead to very low resource utilization. All the resources are allocated in advance, but all may not be required by the process at the same time. Hence some resources will be idle, which will result in low utilization of resources. It may also lead to starvation of processes, as a process may need some very popular resources for its execution. It may not get those resources because of their constant usage by other processes.

No Preemption: This condition can be prevented from occurring in many ways. If a process is requesting a resource that is busy, then all the resources held by the requesting process should be preempted. Another way is to preempt some of its resources, but not all. Preempt only those resources that are required by other processes. In this way, the no preemption condition can be prevented.

But preventing this condition is also not good, because it will result in a wastage of time, and the process already waited in grabbing the resources. In addition, the process needs to be restarted, which results in overhead.

Circular Wait: This condition can be prevented by imposing a complete ordering in which a process should request resources. The process should be allowed to request resources in such a way so that a cycle is not formed. All the resources are numbered, and the process can only request resources in increasing number order. The main aim of doing this is to prevent the formation of a cycle.

But it is not always possible to follow this rule. Another problem is determining how to do the relative ordering of the resources.

⊙ Self-Quiz

Q7. Name three methods that can be used to handle deadlock.
Q8. How can deadlock prevention be done?
Q9. What is mutual exclusion?
Q10. Give one situation where there is no need of mutual exclusion in a computer system.
Q11. What is the drawback of preempting resources from the processes?

DEADLOCK AVOIDANCE

In the deadlock prevention method as explained in Section 7.4, one of the four conditions needs to be prevented, which is not always possible. Hence another method called deadlock avoidance is used. In this method, deadlock is avoided. Whenever any process requests a resource, the system state is checked. If granting the request leaves the system in a safe state, then the resource is allocated; otherwise, it is not. A system is considered to be in a safe state if all the requests of processes are granted to acquire the various resources without leading the system into a deadlock state. A system is considered safe if there is a safe sequence of processes.

- A safe sequence of processes is {P0, P1, P2, . . . Pn} such that all of the resource requests for P_i can be granted using the resources currently allocated to P_i and all processes P_j where j < i.
- If all the processes prior to P_i finish and free up their resources, then P_i will be able to finish also, using the resources that have been freed up.
- If a safe sequence does not exist, then the system is in an unsafe state, which may result in deadlock.
- All safe states are free from deadlock, but unsafe states may result in deadlocks.

This means whenever a request comes from a process for a resource, the whole system is checked for a safe or unsafe state. Two different situations are possible in a system, which are as follows:

- Systems having a single instance of resources.
- Systems having multiple instances of resources.

Separate methods are used in each situation for deadlock avoidance.

Systems Having a Single Instance of Resources

A system having a single instance means that there is just one copy of the particular resource available. In such a situation, a resource allocation graph algorithm is used as a solution in deadlock avoidance.

Resource Allocation Graph Algorithm

If there are only single instances of the resources, then deadlock can be detected by the presence of cycles in the resource allocation graphs. Little modifications are done in the RAG. Now along with the request edges and assignment edges, one more edge called the claim edge is used. The claim edge is a directed edge. It is represented by a dashed line. This is shown in Figure 7.5.

A claim edge points from a process to a resource that it may request in the future. In this algorithm, all claim edges are added to the graph for any particular process before that process is allowed to request any resources. Processes may only make requests for resources for which they have already established claim edges. When a process makes a request, the claim edge $P_i \rightarrow R_j$ is converted to a request edge. This is shown is Figure 7.6.

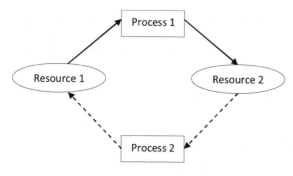

FIGURE 7.5. Resource Allocation Graph (RAG) for deadlock avoidance

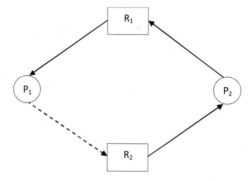

FIGURE 7.6. An unsafe state in a Resource Allocation Graph

When a resource is released, the assignment edge converts back to a claim edge. If resource allocation results in a cycle, then the request will not be granted even if the resource is available. This approach works by denying requests that would produce cycles in the resource allocation graph, taking claim edges into effect.

Systems Having Multiple Instances of Resources

When more than one instance of a resource is present, the resource allocation graph method is not successful. Hence another method called the Banker's algorithm is used.

Banker's Algorithm

The Banker's algorithm gets its name from the basic concept of the banking system. A banker will never loan out money to its clients until it is assured that the loan will be enough to satisfy the needs of all customers. Whenever a process starts it will declare its maximum requirement of resources in advance. Whenever any request is made from any process, before granting the request the scheduler will check whether it will leave the system in a safe state. If the system goes into an unsafe state, then the process will have to wait even if resources are available. The Banker's algorithm uses four main data structures, where P is the number of processes and R is the number of resource types:

- `Available[R]` represents how many resources are presently available of each type.
- `Max [P] [R]` represents the maximum demand of each process of each resource.
- `Allocation [P] [R]` represents the number of each resource type allocated to each process.
- `Need [P] [R]` represents the remaining resources needed of each type for each process.
- `Need[ i ][ j ] = Max[ i ][ j ] - Allocation[ i ][ j ]` for all i, j

Safety Algorithm

Before applying the Banker's algorithm, some method is required to determine whether the system is in a safe state or not. Hence an algorithm is suggested to find the state of the system, and it is called the safety algorithm. This algorithm is used to determine if the present state of a system is safe or not.

The following steps are used in the safety algorithm:

1. Let Work and Finish be vectors of length *m* and *n* respectively.

   - Work is a working copy of the available resources, which will be modified during the analysis.
   - Finish is a Boolean vector indicating whether a particular process can finish or has finished so far in the analysis.
   - Initialize Work to available, and Finish to false for all elements.

2. Find an i such that both (A) `Finish[ i ] == false`, and (B) `Need[ i ] < Work`. This process has not finished, but could with the given available working set. If no such i exists, go to Step 4.
3. Set `Work = Work + Allocation[ i ]`, and set `Finish[ i ]` to true. This corresponds to process i finishing up and releasing its resources back into the work pool. Then loop back to Step 2.
4. If `finish[ i ] == true` for all i, then the state is a safe state, because a safe sequence has been found.

Resource-Request Algorithm (The Banker's Algorithm)

This algorithm is used to determine if a new request is safe. The request will be granted if it leaves the system in a safe state. The request should always be less than or equal to presently available resources. If the above written statement is true, then pretend that it has been granted. After that, check if the resulting state is safe or not.

If the system is safe, grant the request, and if not, deny the request, as follows:

1. Let Request [P] [R] indicate the number of resources of each type currently requested by processes. If Request [i] > Need[i] for any process i, raise an error condition.
2. If Request [i] > Available for any process i, then that process must wait for resources to become available. Otherwise the process can continue to Step 3.
3. Check to see if the request can be granted safely by pretending it has been granted and then seeing if the resulting state is safe. If so, grant the request, and if not, then the process must wait until its request can be granted safely. The procedure for granting a request (or pretending to for testing purposes) is:

   ■ Available = Available - Request
   ■ Allocation = Allocation + Request
   ■ Need = Need – Request

Example 7.1: Consider a system consisting of five resources of the same type which are to be shared by four processes. Each process needs at most two resources. Show that the system is deadlock free.

Solution: If the system is deadlocked, it implies that each process is holding one resource and is waiting for one more. Since there are four processes and five resources, one process must be able to obtain two resources. This process requires no more resources and therefore it will return its resources when done.

Example 7.2: Consider a system having five processes from P0 to P5. There are three types of resources having various instances. Type A has eleven instances, type B has six, and type C has seven instances. A snapshot of the system is shown in Table 7.1.

TABLE 7.1. Processes and Resources Using Various Instances

Process	Allocation			Max			Available		
	A	B	C	A	B	C	A	B	C
P_0	1	2	1	8	6	4	4	4	3
P_1	3	1	1	4	3	3			
P_2	4	1	3	10	1	3			
P_3	3	2	2	3	3	3			
P_4	1	1	3	5	4	4			

Answer the following questions using the banker's algorithm:

a. What is the content of the matrix need?
b. Is the system in a safe state?
c. If a request from process P1 arrives for (1, 0, 2), can the request be granted immediately?

Solution:

a. Need = Max – allocation

Applying this formula to Table 7.1, the matrix *Need* will be as shown in Table 7.2.

TABLE 7.2. Need Matrix

	Need		
	A	B	C
P_0	7	4	3
P_1	1	2	2
P_2	6	0	0
P_3	0	1	1
P_4	4	3	1

b. After applying the banker's algorithm for finding the safe state, the sequence `<P1, P3, P4, P2, P0>` is found. As the algorithm results in a safe sequence, the given system is in a safe state. This means that there is no deadlock right now.

c. Now a request from process P1 arrives for (1, 0, 2). It is to be found out whether this request may be granted immediately or not. For this the request should be less than or equal to availability, meaning `request <= availability`. For process P1, `request (1,0,2) <=availability (3,3,2)`. Hence the request can be granted on a temporary basis. On granting the request on a temporary basis, a new matrix will be formed, which is shown in Table 7.3.

TABLE 7.3. Snapshot of System after Granting Request Temporarily

Process	*Allocation*			*Need*			*Available*		
	A	B	C	A	B	C	A	B	C
P_0	1	2	1	7	4	3	2	3	0
P_1	4	1	3	0	2	0			
P_2	4	1	3	6	0	0			
P_3	3	2	2	0	1	1			
P_4	1	1	3	4	3	1			

Now again the banker's algorithm is applied to find a safe sequence. After applying the banker's algorithm, a sequence `<P1, P3, P4, P0, P2>` is found, which indicates that the system is in a safe state. When the system is in a safe state, the request of process C(1,0,2) can be granted immediately.

(•) Self-Quiz

Q12. What is meant by the term deadlock avoidance?

Q13. Which method is used in deadlock avoidance when a single instance of resources is present?

Q14. Which method is used in deadlock avoidance when multiple instances of resources are present?

Q15. How did the banker's algorithm get its name?

DEADLOCK DETECTION

Deadlock will never occur in the deadlock avoidance method, but the drawback is it needs a lot of calculations before fulfilling any request of the process. These extensive calculations are a burden on the CPU, because during that time no process can occur. In the deadlock avoidance method, before any request may be granted, a safe sequence must to be found. Because of this drawback, some other and better method is required. The suggested method is deadlock detection. The following is true with this method:

■ Deadlocks are allowed to occur in the system.
■ They are detected from time to time.
■ Recovery methods are used to recover from deadlock.

Two different cases are possible in the deadlock detection method, as discussed in the following section.

Systems Having a Single Instance of Resources

If the system has only a single instance of each resource type, a wait-for graph is used. A wait-for graph is a variation of the resource allocation graph. A wait-for graph can be made from a resource allocation graph by removing the resources and edges. An edge from P_i to P_j in a wait-for graph indicates that process P_i is waiting for a resource that process P_j is currently holding. If cycles are present then it indicates the presence of deadlocks. Wait-for graphs are maintained in the system. From time to time, the presence of cycles is searched for, which helps in the detection of deadlock.

Systems Having Multiple Instances of Resources

The detection algorithm is used to detect the presence of deadlocks. It is similar to the banker's algorithm explained in Section 7.5.2.1. But there are two main differences between the detection algorithm and the banker's algorithm:

■ In Step 1, the banker's algorithm sets Finish[i] to false for all i. The detection algorithm sets Finish[i] to false only if Allocation[i] is not zero. If the currently allocated resources for this process are zero, the algorithm sets Finish[i] to true. This is essentially assuming that IF all of the other processes can finish, then this process can finish also. Furthermore, this algorithm is specifically looking for which processes are involved in a deadlock situation, and a process that does not have any resources allocated cannot be involved in a deadlock, and so can be removed from any further consideration.

■ Steps 2 and 3 are unchanged.

■ In Step 4, the basic banker's algorithm says that if Finish[i] == true for all i, there is no deadlock. The detection algorithm is more specific, by stating that if Finish[i] == false for any process P_i, then that process is specifically involved in the deadlock that has been detected.

Detection-Algorithm Usage

In the deadlock detection method, the main question is when deadlock detection should be done. The answer may depend on how frequently deadlocks are expected to occur. There are two basic approaches. In the first approach deadlock detection is done after every resource allocation. But this approach has a lot of calculation overhead on the CPU. In the second approach deadlock detection is done when CPU utilization reduces below the desired value.

◉ Self-Quiz

Q16. What is the main drawback of the deadlock avoidance method for handling deadlock?

Q17. Deadlocks are allowed to occur in the deadlock detection method. Is this statement true or false?

Q18. Name the type of graph that is maintained in the deadlock detection method.

Q19. How is a wait-for graph different from a resource allocation graph?

RECOVERY FROM DEADLOCK

There are three basic ways to recover from deadlock:

1. Inform the system operator, and deadlock is removed manually.
2. Terminate one or more processes involved in the deadlock.
3. By preemption of resources.

Process Termination

Terminate all processes which are involved in the deadlock. This approach will resolve the deadlock, but in this case more processes are terminated than required. Terminate processes one by one until the deadlock is broken. In this way a lower number of processes will be terminated. However, deadlock detection is required after each step. When processes are to be terminated one by one, then the victim process is to be chosen. The victim process is chosen based on the following points:

1. Priority of the process.
2. Total processing done of the particular process.
3. How close a process is to its finish.
4. How many resources a process is holding.
5. Which types of resources a process is holding.
6. How many more resources the process needs to complete.
7. Whether the process is interactive or batch.

Resource Preemption

When preempting resources to relieve deadlock, there are three important issues to be taken care of.

1. **Selecting a victim:** To choose out of many processes a victim process to be terminated first.
2. **Rollback:** It is ideal to roll back a preempted process to a safe state in place of terminating it fully.
3. **Starvation:** How do you guarantee that a process won't starve because its resources are constantly being preempted? One option would be to use a priority system and increase the priority of a process every time its resources get preempted. Eventually it should get a high enough priority that it won't get preempted anymore.

SUMMARY

Various methods and ways are used in an operating system to deal with the problem of deadlock. Approaches such as deadlock prevention, deadlock avoidance, deadlock detection, and recovery have been explained in the chapter.

POINTS TO REMEMBER

- A situation when two or more processes are blocked and cannot proceed further with their processing because of interdependency of resources is called deadlock.
- In a computer system, deadlock arises when four conditions occur simultaneously.
- The four conditions are mutual exclusion, hold and wait, no preemption, and circular wait.
- A directed graph which consists of two types of nodes as processes and resources is called a resource allocation graph.
- Resource allocation graph in short is also known as RAG.
- The presence of deadlock can be indicated with the help of RAG.

- The main methods to deal with deadlock are deadlock prevention, deadlock avoidance, and deadlock detection.
- In the deadlock detection method, deadlocks are allowed to occur in the system. They are detected from time to time.
- Recovery methods are used to recover from deadlock.

REVIEW QUESTIONS

Q1. Explain different conditions for deadlock.

Q2. Write down the methods for deadlock prevention.

Q3. How is recovery from deadlock done by using a combined approach?

Q4. (i) State four necessary conditions for deadlock to occur. Give a brief argument for the reason each individual condition is necessary.

(ii) In the deadlock prevention method, the system grants resources on an "all or none" basis to processes. Discuss its pros and cons.

Q5. Differentiate between deadlock and starvation.

Q6. Write and explain the banker's algorithm for avoidance of deadlock.

Q7. Write an algorithm for the detection of deadlock in a system having several instances of multiple resource type.

CHAPTER **8**

MAIN MEMORY MANAGEMENT

Purpose of this Chapter

The purpose of this chapter is to explain the need of main memory management. Various types of memory management schemes are explained. The main purpose of all the techniques is to allocate each program and its required memory space. Concepts like paging, segmentation, and segmented paging are also explained in detail.

INTRODUCTION

Whenever the programs are to be executed, they should be present in the main memory. In multiprogramming, many programs are present in the main memory, but the capacity of memory is limited. All of the programs cannot be present in memory at the same time. This memory management is to be performed so all of the programs get space in memory, and they get executed from time to time. Various memory management schemes are there. The selection of a particular scheme depends on many factors, especially the hardware design of the system.

MAIN MEMORY

Main memory is the temporary read/write memory of a computer. It is a set of continuous locations. Each location has a specific address. Each address is a binary address. For the convenience of the programmer and user, they are represented in hexadecimal numbers.

The programs, which are a set of instructions, are stored in secondary memory. When they need to be executed, they are loaded to main memory from the secondary memory. The main memory instructions are stored in various locations according to space availability.

The groups of memory locations are called the *address space*. Two types of address space are explained below, i.e., *physical address space* and *logical address space*.

Physical Address Space

The address of any location in physical memory is called the *physical address*. Physical memory means the main memory or Random Access Memory (RAM). In main memory every location has an address. Whenever any data or information is read from it, its address is given. That address is called the physical address.

A group of many physical addresses is called the *physical address space*. The Features of the physical address space follow:

- The physical address is provided by the hardware.
- It is a binary number made with a combination of 0 and 1.
- It refers to a particular cell or location of primary memory.
- Such addresses have some end limits.
- The limits normally start from zero to some end limit (shown in Figure 8.1).
- All addresses do not only belong to the system's main memory.

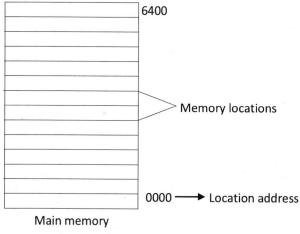

Main memory

Main memory having locations from 0000-6400

FIGURE 8.1. Main memory having locations from 0000 to 6400

Logical Address Space

The address of any location of virtual memory is called the *logical address*. It is also known as the *virtual address*.

The group of many logical addresses is called the *logical address space* or *virtual address space*. The features of the logical address space follow:

- Logical address is provided by the operating system kernel.
- Logical address space may not be continuous in memory.
- It might also be present in the form of segments.
- Sometimes logical addresses might be of the same value as a physical address.

- The logical address is mapped to get the physical address.
- This mapping is done using address translation.
- The physical address space and logical address space are independent of each other.

Mapping from the Logical Address to the Physical Address

The memory management unit is used to map the logical address to the physical address. The memory management unit is also known as *MMU*. It consists of a relocation register.

The logical address can be converted into a physical address using the relocation register as shown in Figure 8.1. This register consists of a value.

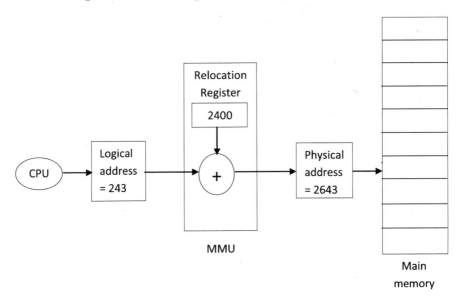

FIGURE 8.2. Main Memory Management Unit (MMU)

- The CPU generates a logical address when a process wants to address a location.
- The value of the relocation register is added to the logical address.
- The new value which is produced as output is the value of a physical address.
- This physical address will point to a location in main memory.
- This location will be used as a pointer to the main memory, where the operation is to be performed.
- Mapping of the virtual address into a physical address is also known as *address translation* or *address binding*.

⊙ Self-Quiz

Q1. Define physical address and logical address.
Q2. What is the difference between the physical address space and the logical address space?

OVERLAYS

The process is brought into the main memory to be executed. Sometimes the process size is large when compared to the available free space in the main memory. A technique is used to help the process that is to be executed, even if it does not have enough space in memory. This technique is called *overlays*.

- With this technique, the complete process is *not* kept in the main memory.
- Only some instructions and data required by the process are kept.
- Only those instructions that are to be executed *first* are placed in the main memory.
- The rest of the process will be in disk.
- When other instructions are executed, they are placed in main memory.
- At that time, the previous instructions are removed, and space is created for the next set of instructions.
- In this way a process whose size is much larger than the available memory can be executed.
- This technique of overlays is implemented by the user.

An example of a two pass assembler is considered. Its different parts and their size requirements are given in Table 8.1. Its total memory requirement is 200 KB, whereas the available memory is only 150 KB.

TABLE 8.1. Parts of two pass assembler and their memory requirements

Content	Size
Pass 1	60 KB
Pass 2	90 KB
Symbol Table	20 KB
Routines	30 KB

Both Pass 1 and Pass 2 are not required to be present at the same time in memory for the execution of the two pass assembler.

- First Pass 1 is loaded into memory.
- Its function is to construct a symbol table.
- When its work finishes, then it is removed from the main memory.
- When completed, Pass 2 is loaded into main memory as shown in Figure 8.3.
- Its work is to generate the machine language code.
- This removal and bringing of required code into the memory is done by overlay drivers.
- The overlay driver is loaded into memory.

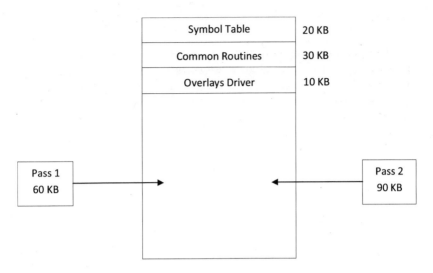

FIGURE 8.3. Overlays for a Two Pass Assembler

SWAPPING

When many processes are going to be executed and there is not sufficient space in the main memory, then a method called *swapping* is used. In this method, a lower priority process is removed from the main memory on a temporary basis to create space for a higher priority process.

■ A process is swapped out of main memory on a temporary basis.
■ The swapped process is placed on the disk. This is also known as a *backing store*.

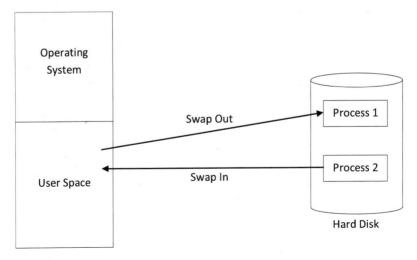

FIGURE 8.4. Swapping of processes

- The higher priority process is placed in the place of the swapped process.
- Lower priority processes are swapped out of memory.
- Higher priority processes are swapped into memory.
- This method is also known as *Roll In and Roll Out* and shown in Figure 8.4.
- The overhead in this method is the swap time.
- This technique is used in many operating systems such as UNIX, Windows, and Linux, etc.

Self-Quiz

Q3. Define overlays.
Q4. Why overlays are used?
Q5. What do you mean by swapping?

FIXED PARTITION MEMORY MANAGEMENT

In this memory management technique, the primary memory is divided into fixed partitions. Dividing the main memory into sets of regions which are non-overlapping is called *partitioning*. These partitions can be of equal or unequal sizes as shown in Figure 8.5. The size of the partition is fixed at the time of OS initialization and cannot be changed at runtime of programs. This is the reason it is called *fixed partitioning memory management*.

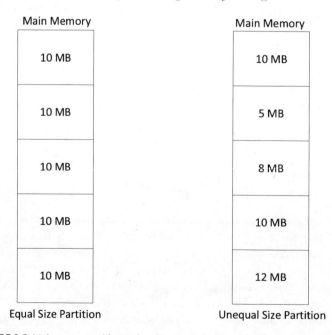

FIGURE 8.5. Main memory with equal and unequal sized partitions

Any program or process that is smaller than or equal to a partition can be loaded into the free partition. If there is no free partition, the operating system swaps a process out of a partition to make space for it. Sometimes a program may be too large to fit into a partition. In this case, the programmer would use the technique *overlays* as explained in Section 8.2.

Process Allocation in Partitions

- A single queue is maintained.
- The process from that queue is allotted in a free partition in First In First Out (FIFO) manner.

Limitations of Fixed Partitions

- The program size is limited to the largest partition which is available in the main memory.
- This scheme suffers from internal fragmentation. *Internal fragmentation* means the unused space within partitions after the allocation of process in a partition.

VARIABLE PARTITION MEMORY MANAGEMENT

In this memory management scheme, main memory is divided into partitions of variable sizes. The method explained in Section 8.4 suffers from internal fragmentation. The *variable partition memory management scheme* is free from internal fragmentation because the process is allocated to a partition of its size. Memory is not partitioned in advance. When a process is introduced, then only according to its size and available free memory, is it allocated memory space. Memory can be considered as a sequence of variable size free partitions.

Fragmentation in Detail

When a process is terminated, it must be removed from the main memory. This creates free memory space. Some parts of memory are occupied by processes and some locations are free. Such a snapshot of memory is said to be *fragmented*. There are mainly two types of fragmentation, i.e., *external* and *internal*.

- **External Fragmentation:** External fragmentation means memory is available as per the size of the process, but it is not continuous. Total space available in the memory is large enough to accommodate the process, but free space is not continuous.

  This type of fragmentation is explained with the help of an example. Suppose the main memory is allocated to three processes A, B, and C, respectively. There are three non-contiguous holes of free memory spaces of size 2 MB, 6 MB, and 4 MB. This means a total of 12 MB of memory is free, but it is present in fragmented order. In this situation if a new process is loaded into the memory that requires 10 MB of memory space, it cannot be allocated memory because there is no contiguous 10 MB of free memory space available. This snapshot is further shown in the Figure 8.6.

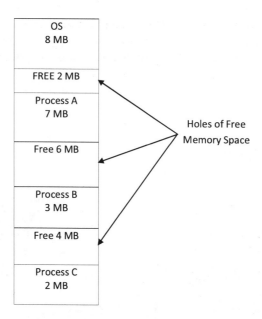

FIGURE 8.6. External Fragmentation

■ **Internal Fragmentation:** Sometimes more memory space is allocated to the process that is actually required by it. This results in a waste of space, as it cannot be allocated to other process. Such fragmentation is called *internal fragmentation*.

Suppose the size of main memory is 64 MB. It is divided into eight equal blocks of 8 MB each. A process will be allocated one or more blocks according to its size. A block cannot be shared between two processes. The processes A and C require less space than 8 MB. One block has to be allocated to each of them. This causes unused memory space within the allotted memory of the processes. This fragmentation is internal fragmentation, which is shown in Figure 8.7.

Compaction

Compaction is bringing free space together into one place in order that free memory is available in a contiguous manner. When two adjacent holes appear then they can be merged together to form a single big hole. This larger hole can be used by the process that needs large memory requirements. Another method is to relocate the processes, thus creating free space in a contiguous manner.

Compaction is used randomly, as free space will be created as soon as processes are terminated and leave the memory. The process of compaction is shown in Figure 8.8.

Process Allocation in Partitions

When a process arrives, then free memory space large enough to accommodate the incoming process is sought. If enough memory is available in a continuous manner, then the incoming process is accommodated in it. Free space is scattered throughout the memory. Three different techniques are used to satisfy the request which are *First fit, Best fit,* and *Worst fit.* The free spaces are called holes. A list of free holes as well as filled memory space is maintained by the operating system.

Block
Numbers

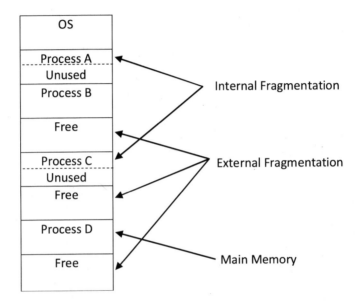

FIGURE 8.7. Internal Fragmentation

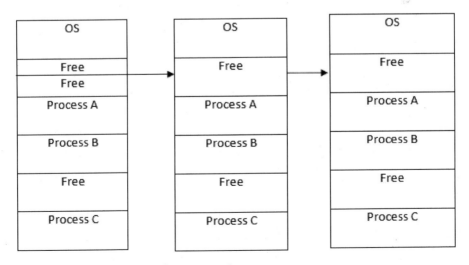

FIGURE 8.8. Compaction of Free Memory Spaces

■ **First fit:** Allocates the first hole which is large enough to satisfy the request. A list of free holes is maintained. Whenever a process arrives, the list of free holes is searched in order to find the first hole large enough to accommodate the incoming process.

■ **Best fit:** Allocates the smallest hole that is large enough to accommodate the incoming process.

■ **Worst fit:** Allocates the largest hole that can accommodate the incoming process. The list of free holes is maintained according to size. The largest hole, in which the incoming process can accept, will be allocated.

Example 8.1: Suppose there are memory partitions of 200 KB, 600 KB, 300 KB, 400 KB and 700 KB. These partitions are in order. Use first fit, best fit, and worst fit algorithms to allocate processes of parameters 312 KB, 517 KB, 212 KB, and 526 KB, and solve the following:

    a. Represent the allocations with the help of a figure or table.

    b. Which algorithms of the three make the most efficient use of memory?

Solution:

    a. Table 8.2 represents the allocations of the processes according to first fit. In this example the process P4 of size 526 KB will not find any free partition to be accommodated in the memory.

TABLE 8.2. First Fit

First Fit		
Process	*Process Size*	*Block*
P$_1$	312 KB	600 KB
P$_2$	517 KB	700 KB
P$_3$	212 KB	200 KB
P$_4$	526 KB	—

Table 8.3 represents the allocations of the processes according to best fit. In this example all processes will get free partitions to be accommodated into the memory.

TABLE 8.3. Best Fit

Best Fit		
Process	*Process Size*	*Block*
P$_1$	312 KB	400 KB
P$_2$	517 KB	600 KB
P$_3$	212 KB	300 KB
P$_4$	526 KB	700 KB

Table 8.4 represents the allocations of the processes according to worst fit. In this example, the process P4 of size 526 kb will not find any free partition to be accommodated in the memory.

a. The best fit algorithm makes the most efficient use of memory as all of the four processes receive space in memory.

TABLE 8.4. Worst Fit

Worst Fit		
Process	*Process Size*	*Block*
P$_1$	312 KB	400 KB
P$_2$	517 KB	600 KB
P$_3$	212 KB	300 KB
P$_4$	526 KB	700 KB

Limitations of Variable Partitions

This scheme suffers from *external fragmentation*. External fragmentation means memory is available according to the size of the process, but it is not continuous. When processes are allocated and de-allocated from memory, they result in scattered partitions.

These free partitions are scattered throughout the memory.

(●) Self-Quiz

Q6. How are processes allocated in memory using the fixed partition method?
Q7. What is the difference between fixed and variable partition memory management schemes?

PAGING

In the *paging* technique of memory management, the physical memory is divided into fixed-sized blocks. Paging is a method of non-contiguous memory allocation. In paging, the logical address space is divided into fixed-sized blocks known as *pages*. The physical address space is also divided into fixed-sized blocks known as *frames*. Every page is must be mapped to a frame. The concept of paging is shown in Figure 8.9. Characteristics include:

■ Divides logical memory into blocks of the same size called pages.
■ The size of the frames is usually in the power of 2.
■ The size of frame usually ranges between 512 bytes and 8192 bytes.
■ Management technique has to keep track of all free frames.
■ If a program needs *n* pages of memory, then *n* free frames are found in the memory and the program is loaded there.
■ Pages are scattered throughout the memory.

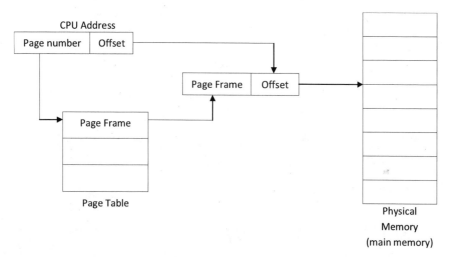

FIGURE 8.9. Paging

- A page table is maintained to translate logical addresses into physical addresses.
- Paging suffers from internal fragmentation.
- Paging does not suffer from external fragmentation.

Did you know?	The Linux operating system uses a page size of 4 KB.

The address generated by the CPU is divided into two parts: *page number* and *page offset*.

- Page number (p) – used as an index into a page table that contains the base address of each page in physical memory.
- Page offset (d) – combined with the base address to define the physical memory address that is sent to the memory unit.

Mapping of Pages to Frames

For example, suppose the logical address space has 2^{13} addresses. This means that a logical address has 13 bits. The logical address space is 8 KB. There are 8 pages in logical address space. The size of each page will be 2^{10} bits which means 1 KB. The size of physical address space is 2^{12} bits which means 4 KB. This means that a physical address is 12 bits long. In a logical address of 13 bits the Most Significant Bit (MSB) (13-10) = 3 bits represents page number and the rest of the bits represent the offset. If the logical address is 011 1010101010, then it denotes the page number 011 (page 3) and offset 1010101010.

The operating system maintains a table to convert logical address into physical address. This table is called a page table. The method of converting the logical address into a physical address is called *address translation*. The process of address translation for address 011 101010101 is explained with the help of Figure 8.10.

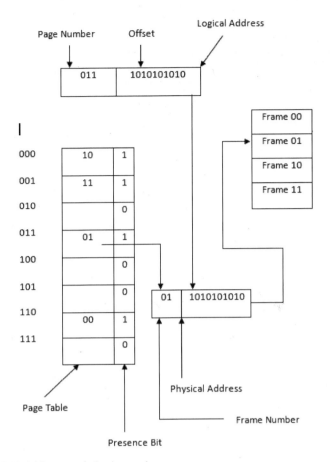

FIGURE 8.10. Address translation in a paging system

Page Table

The page table is maintained in main memory. The address of the page table in main memory is stored in a register which is called the *Page Table Base Register (PTBR)*. The size of page table is stored in a register called the *Page Table Length Register (PTLR)*. In this method of paging, the page table two memory accesses are used whenever any data is to be retrieved from main memory. One memory access is for age memory and a second memory access is from main memory. Two memory accesses for any data retrieval use great amounts of time. This problem is removed by using the *Translation Look Aside Buffer Register (TLBR)* as shown in Figure 8.11.

Advantages of Paging

- Paging does not suffer from external fragmentation.
- Memory management algorithm is easy to use.
- As pages and frames are of equal size, swapping is easy.

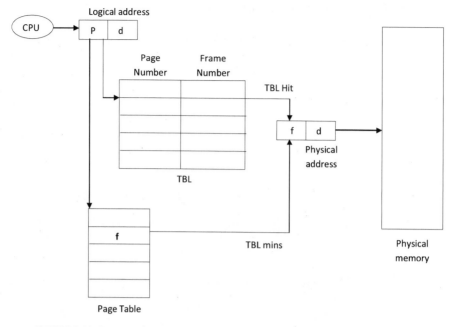

FIGURE 8.11. Concept of paging with Translation Look Aside Buffer (TLB)

Limitations of Paging

- Paging suffers from internal fragmentation.
- Additional memory space is required to store the page table, as the page table is maintained in main memory.
- Multilevel paging increases number of memory references.

Did you know?	The Linux operating system uses a three level page table.

⊙ Self-Quiz

Q8. Define paging.
Q9. Why is page size normally in the power of 2?
Q10. Paging suffers from which type of fragmentation?
Q11. Where is page table maintained?
Q12. What is the purpose of a page table?

SEGMENTATION

Segmentation is a technique of non-contiguous memory allocation. In this method memory space is divided into unequal sized chunks which are known as *segments*. In segmentation, an application is considered as a collection of modules where each module can be loaded into a different segment. For example an application may consist of many modules like the main program, procedures, functions, symbol tables, data values, global variables, local variables, etc., and each module can be placed in a different segment. This can be represented with the help of Figure 8.12, where 1-6 are different modules of any application.

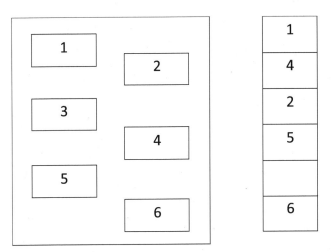

FIGURE 8.12. User view of memory

The features of segmentation follow:

- This method supports user's view of memory.
- Logical address is divided in two parts: segment number and offset.
- Segment table is maintained.
- Each entry of the table has two values: base and limit.
- *Base* is the starting physical address where the segments reside in memory.
- *Segment Table Base Register (STBR)* points to the segment table's location in memory.
- *Segment Table Length Register (STLR)* indicates number of segments used by a program.
- Segment number s is legal if $s <$ STLR.

Implementation of Segmentation Hardware

Implementation of the segmentation technique is shown in Figure 8.13.

Accessing Addresses in Segmentation

For example the logical address 310000 indicates offset 10000 and segment number 3. A segment table is used for validation of an address. The starting address and size of the segments are stored in the segment table. It is represented with the help of an example shown in Figure 8.14.

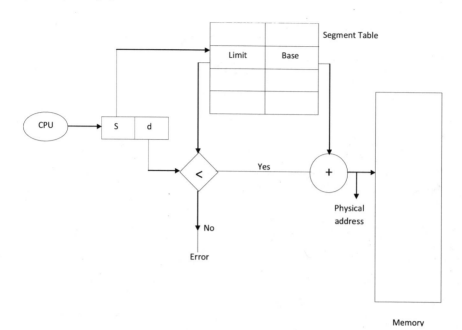

FIGURE 8.13. Segmentation

Segment number	Base Address	Segment Size
0	75000	5000
1	20000	1000
2	60000	15000
3	21000	35000
4	80000	10000

Segment Table

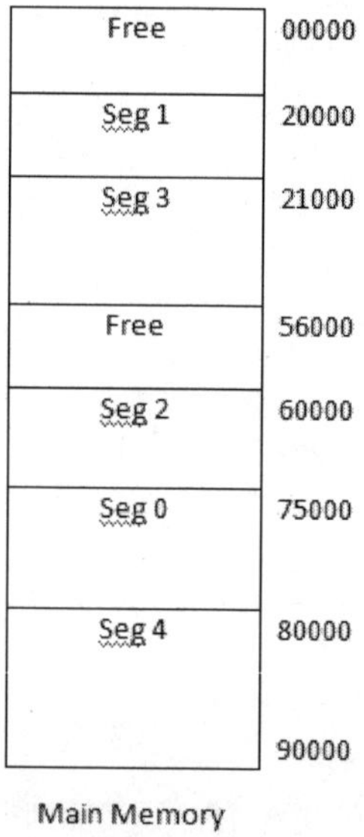

Free	00000
Seg 1	20000
Seg 3	21000
Free	56000
Seg 2	60000
Seg 0	75000
Seg 4	80000
	90000

Main Memory

FIGURE 8.14. Segments and segment table

Advantages of Segmentation

- Segmentation does not suffer from internal fragmentation.
- The segment table consumes less memory space than page tables.
- As segment tables are small, memory reference is easy.

Limitations of Segmentation

- Memory management algorithm used is complex.
- It suffers from external fragmentation.
- Swapping is not easy as segments are of unequal sizes.

(•) Self-Quiz

Q13. Define segmentation.
Q14. Which type of fragmentation is present in segmentation?
Q15. What are the different parts of the logical address in segmentation?

SEGMENTED PAGING

The techniques of memory management explained in Sections 8.6 and 8.7, i.e., paging and segmentation, have their advantages and disadvantages. Thus, the advantages of both are taken and a new approach of memory management is suggested which is called *segmented paging*. In this technique the user's memory space is broken into segments. Each segment is further broken into fixed size pages as shown in Figure 8.15. Segmentation is visible to the user, whereas paging is transparent to the user.

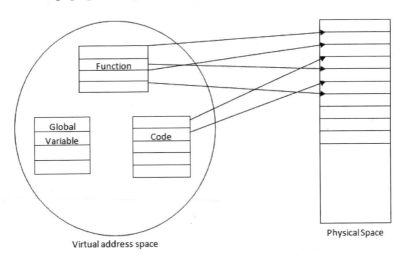

FIGURE 8.15. Segmented Paging

Address Translation in Segmented Paging

In this method two tables are used, i.e., a segment table and a page table. The segment table has two fields: the base address and the size of segment. The base address of a segment is the address of the first page in each segment. The size of the segment is the number of the pages present in each segment. The page table stores the address of the frame where a page is loaded into memory. The logical address has three parts: segment number, page number, and an offset.

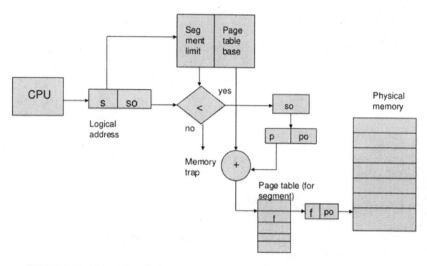

FIGURE 8.16. Address Translation in Segmented Paging

Example of Segmented Paging

Below is an example of segmented paging:

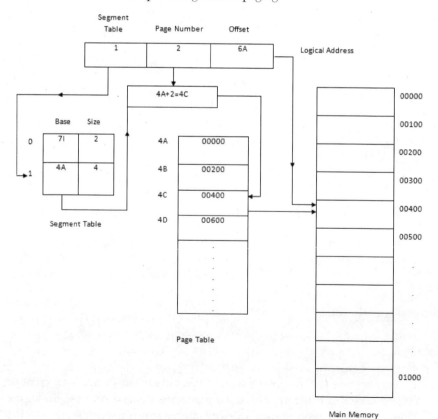

Advantages of Segmented Paging

- Memory space used is less as compared to the pure paging technique.
- Individual pages are shared by copying page table entries.
- External fragmentation is not present in this method.
- Memory allocation is simple.
- *Intel Pentium* supports the memory management techniques of segmented paging.

(⊙) Self-Quiz

Q16. Define segmented paging?
Q17. What are the advantages of segmented paging compared to pure segmentation and pure paging techniques?

PROTECTION SCHEMES

In almost all modern operating systems, for example Linux, Windows, etc, it is normal to have many programs running concurrently in memory. Although there are advantages, there are many problems with this. One problem is to protect programs from interfering with one another. This is called *protection*. Protection in the OS is necessary so processes do not refer to memory locations of another process without permission. Address references must be checked at runtime by the hardware. Protection in paging is suggested and explained in the next section.

Protection in Paging

In a paged environment, memory protection is done by *protection bits*. These bits are associated with each frame. Usually, these bits are stored in a page table. One bit is used to define a page which is read-write or read-only. Whenever data is retrieved from memory, the request will go to the page table to find the correct frame number. Simultaneously, the physical address is calculated. The protection bits are checked to verify that no writes are being made to a read-only page. An attempt to write to a read-only page causes a hardware trap to the operating system, and it results in memory-protection violation. One additional bit named as a **valid-invalid** bit is attached with each entry in the page table.

- When the valid-invalid bit is set to 1, then the page is considered as valid.
- When the bit is set to 0, then the page is invalid. This indicates that the requested page is not in the logical address space of the process.

The OS sets this bit for each page which helps in deciding to grant or discard access to the page. Illegal addresses are trapped with the use of the valid-invalid bit. The use of a valid-invalid bit is explained further with the help of an example. Let's suppose there are memory locations ranging from 0 to 16383. For these many locations a total of 14 bits are required. The page size is 2 KB; thus, a total of 6 pages will be present. Pages 0, 1, 2, 3, 4, and 5 (total 6 pages) are mapped through the page table. If any attempt is made to generate an address in pages 6 or 7, then the valid-invalid bit will be designated as invalid, and the computer will trap to the OS with an invalid page reference. See Figure 8.17.

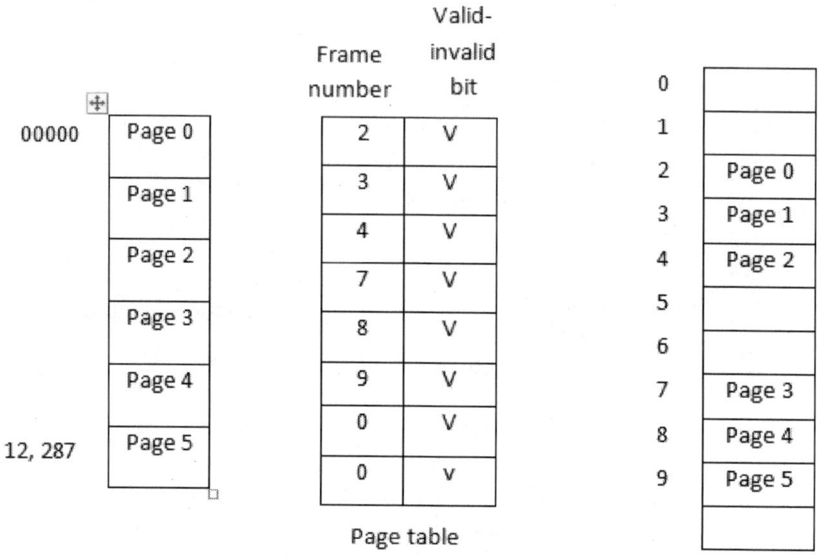

		Frame number	Valid-invalid bit		
00000	Page 0	2	V	0	
	Page 1	3	V	1	
		4	V	2	Page 0
	Page 2	7	V	3	Page 1
	Page 3	8	V	4	Page 2
		9	V	5	
	Page 4	0	V	6	
12, 287	Page 5	0	v	7	Page 3
				8	Page 4
				9	Page 5

Page table

FIGURE 8.17. Valid (v) or invalid (i) bit in a page table.

COMPARISON BETWEEN PAGING AND SEGMENTATION

- In segmentation, the programmer is involved in creating and maintaining the segments, whereas in paging there is no involvement of programmer.
- Segmentation is not transparent as the paging memory management scheme. The programmer is not aware of the presence of paging in the system, whereas this is not the case with segmentation.
- In segmentation, a mechanism provides protection to individual segments, whereas no separate protection mechanism exists in paging.
- Only one linear address space is present in paging, whereas many linear address spaces are present in segmentation.
- Shared code is present in segmentation but it is not in paging. This means that sharing of pages is not allowed in the paging method, whereas sharing of code is allowed in segmentation.
- The concept of paging was designed to get a large linear address space without buying more physical memory, whereas segmentation was designed to allow programs and data to be broken into logically independent address spaces and to provide security and protection to individual segments.

SUMMARY

The main memory stores the instructions which are to be executed by the CPU. The organization and management of memory is the important task of the OS. The OS allocates memory to various processes. Overlays and swapping are explained as memory management schemes. Memory management schemes for fixed partitions and variable partitions are explained. Memory suffers from internal and external fragmentation. Memory allocation schemes, i.e., paging and segmentation are also explained.

POINTS TO REMEMBER

- Physical memory means main memory or RAM.
- The address of any location in physical memory is called the physical address.
- A group of many physical addresses is called the physical address space.
- The address of any location of virtual memory is called the logical address. It is also known as the virtual address.
- A group of many logical addresses is called the logical address space or virtual address space.
- The logical address is to be mapped into the physical address.
- A technique is used with the help of which a process can be executed even if it does not get the complete space in memory. This technique is called overlays.
- Swapping is a method in which a lower priority process is removed from the main memory on a temporary basis in order to create space for a high priority process.
- Paging is a method of non-contiguous memory allocation.
- In paging, the logical address space is divided into fixed sized blocks known as pages.
- The size of the pages is usually in the power of 2.
- Paging suffers from internal fragmentation.
- Paging does not suffer from external fragmentation.
- Segmentation is a technique of non-contiguous memory allocation.
- In segmentation, an application is considered as a collection of modules where each module can be loaded into different segments.
- This method supports the user's view of memory.
- Protection in the OS is necessary so that processes do not refer memory locations of other processes without permission.
- In a paged environment, memory protection is done by *protection bits*.

REVIEW QUESTIONS

Q1. Discuss the paging system for memory management in detail. Also discuss its advantages and disadvantages.

Q2. Discuss the similarities and differences between paging and segmentation.

Q3. What do you understand by fragmentation? Explain each type in detail.

Q4. What are the different techniques to remove fragmentation in case of multiprogramming with variable and fixed partitions? Discuss.

Q5. What is paging? Describe how a logical address is converted into a physical address in a paged system. Further give two reasons as why page sizes are kept in powers of 2.

Q6. Discuss the paged segmentation scheme of memory management, and explain how a logical address is converted into physical address in such a scheme.

Q7. What is paging and segmentation?

Q8. Explain the differences between internal and external fragmentation.

Q9. What is the difference between paging and segmentation?

Q10. Consider a logical address space of eight pages of 1024 words, each mapped onto a memory of 32 frames then:
   a. How many bits are there in logical address?
   b. How many bits are there in physical address?

CHAPTER 9

VIRTUAL MEMORY

Purpose of the Chapter

Sometimes the size of a program is larger than the memory space available to the programmer. To allow the programmer to manage programs larger than the main memory size, the concept of virtual memory is used. This chapter explains various elements of virtual memory, such as page fault, demand paging, page replacement, and its algorithms. It further explains various issues related to performance of virtual memory like allocation of frames and thrashing.

INTRODUCTION

Whenever programs are to be executed, they are placed in main memory. If the size of the program is larger than the available memory, then this situation should not limit the execution of programs. To solve this issue many techniques like overlays and swapping are explained in Chapter 8. Another concept known as virtual memory is also used to solve this issue of main memory space limitation. The basic concept of virtual memory is to store the instructions and data of the program in the secondary memory and bring them into the main memory when they are required for execution. In other words, the part of the program that is required at any moment is loaded in main memory while rest of the program is present in secondary memory. Virtual memory is useful in a multitasking operating system, where memory available in the main memory for user applications is divided among several user programs and every program has to compete for its memory requirements.

A technique that allows the execution of processes that may not be completely in main memory is called virtual memory.

Hence in virtual memory:

- There is a separation of user logical memory from physical memory.
- Only the part of the program required for execution is kept in memory.
- Logical address space is much larger than physical address space.
- Address spaces are allowed to be shared by several processes.

METHODS TO IMPLEMENT VIRTUAL MEMORY

Virtual memory is implemented in ways such as *demand paging* and *demand segmentation*.

- In demand paging, the available memory space is divided into equal-size pages, whereas in demand segmentation, memory is divided into various-sized segments.
- The demand paging method is mainly used to implement virtual memory because segment replacement algorithms are more complex than page replacement algorithms.
- Segment replacement algorithms are more complex than page replacement algorithms (which will be explained in Section 9.4) because segments are of different (variable) sizes.
- Despite its complexity demand segmentation is used in some operating systems. The IBM OS/2 operating system uses demand segmentation.

In this book, we use the technique of demand paging to implement virtual memory.

Self-Quiz

Q1. Define virtual memory.
Q2. What is the purpose of virtual memory?
Q3. Name the ways used to implement virtual memory.
Q4. Why is the demand paging method more popular than demand segmentation?

DEMAND PAGING

A program is initially stored in the secondary memory. When a page in the program is required, it is swapped into the main memory. This is called demand paging, because a page is not loaded into the main memory until it is required. The technique of demand paging is similar to the paging explained in the previous chapter. The only difference between paging and demand paging is that in demand paging, pages are loaded into the main memory according to need (demand).

The paging technique in which pages are loaded into the main memory according to demand or need is called demand paging.

Page Fault

When a page is required, the operating system scans the page table to find out the physical address in the main memory in which the page is loaded. If the page is found, then the operating system continues its task; otherwise, a page fault occurs.

When the desired page is not present in the main memory, this condition is known as a page fault.

A page fault indicates that the required page is not currently present in the main memory. In the concept of virtual memory, it is not necessary that all pages be present in the main memory all the time. In Figure 9.1 the snapshot of memory is shown in which all pages are not present in the main memory.

In Figure 9.1, there are eight pages ranging from 0–7 present in the logical memory. Out of these eight pages, only four pages are present in the physical memory. For the pages that are present in the main memory, the valid-invalid bit is "v."

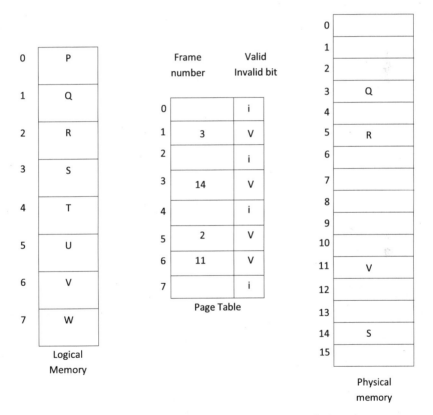

FIGURE 9.1. Snapshot of memory in which all pages are not present in the main memory

Steps to Handle Page Faults

If a process tries to access a page and that page is not present in the main memory, then a page fault occurs. While accessing the page, its page table entry is found to be invalid "I," which indicates that the desired page is not present in the main memory. The situation of a page fault is not desirable, and it results in a page fault trap. There are six main steps to follow for the operating system to handle a page fault trap.

The steps for handling a page fault are as follows, as shown in Figure 9.2:

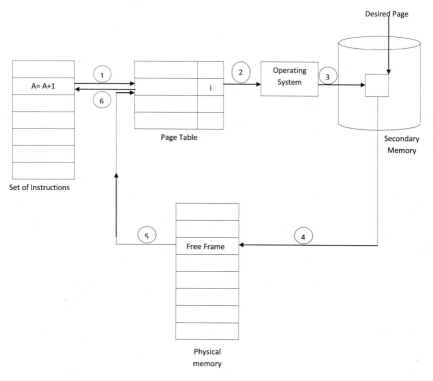

FIGURE 9.2. Steps for handling a page fault

1. The page containing the instruction to be executed is searched in the page table. Then its valid-invalid bit is checked. In a page fault, its entry in the page table will be invalid.
2. When the valid-invalid bit of the desired frame is "I," then a page fault trap occurs and a signal will go to the operating system.
3. The operating system will locate the desired page in secondary memory (backing store).
4. A free frame is located in the main memory. In that free frame the located frame from secondary memory is stored.
5. After storing the desired page in main memory, its page table entry is corrected. Hence the valid-invalid bit is now set to valid "v," as the page is now present in the main memory.
6. The instruction that was interrupted by the illegal page address trap will now be restarted. The process will now get the page into the main memory and the desired instruction will be executed smoothly.

Performance of Demand Paging

Page Fault Rate 0 d" p d" 1.0

If p = 0, there are no page faults

If p = 1, every reference is a fault

Effective Access Time (EAT) = $(1 - p)$ x memory access + p (page fault overhead + [swap page out] + swap page in + restart overhead)

For better performance of demand paging, the effective access time should be as small as possible. The smaller the value of EAT, the better the performance of demand paging.

Example of Demand Paging

Suppose the memory access time is equal to 1 microsecond. Fifty percent of the time, the page that is being replaced has been modified, and hence it needs to be swapped out.

Swap Page Time = 10 msec = 10,000 msec

Effective Access Time (EAT) = $(1 - p)$ x 1 + p (15000)

1 + 15000P (in msec)

Advantages of Demand Paging

- It brings a page into memory only when it is needed.
- Less Input/Output (I/O) is needed if demand paging is used.
- Less memory space is required.
- There is a faster response from processes.
- More users are allowed to execute their programs.

⊙ Self-Quiz

Q5. Define demand paging.

Q6. What is a page fault?

Q7. When does a page fault occur?

Q8. "The value of effective access time in demand paging should be at a minimum for better performance." Justify the statement.

PAGE REPLACEMENT

In handling a page fault, what will happen if there is no free frame? The technique known as page replacement is used. In page replacement, find some page in memory which may not be in use and swap it out to create a free frame in memory as shown in Figure 9.3. Various page replacement algorithms exist for selecting the victim page which needs to be swapped out of the main memory. Some pages may be brought into memory several times according to their requirement. Although there are many page replacement algorithms to select the victim page, the best algorithm is the one which gives the minimum number of page faults.

Selecting a page in memory which may not be in use and swapping it out create a free frame in memory and placing the desired page in that newly created free frame is called page replacement.

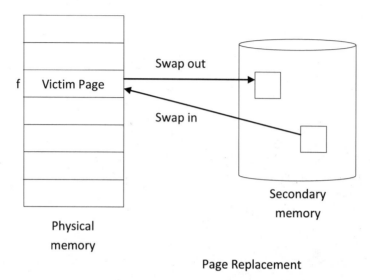

FIGURE 9.3. Page replacement between physical memory and secondary memory

PAGE REPLACEMENT ALGORITHMS

A page replacement algorithm is the algorithm used to select a page which is to be swept out of the main memory to create space for the requested page which has caused the page fault. There are many page algorithms, such as first in first out, optimal page replacement, least recently used, and so on.

First In First Out Algorithm (FIFO)

This is the simplest of all page replacement algorithms. When the page fault occurs, then the oldest page in the memory is swept out of the main memory to create space for the requested page that needs to be executed. A reference string of pages is maintained in memory which is consulted to select a victim. Reference string means that pages are referred from the main memory in this sequence.

In the FIFO algorithm, swap out the page which entered into memory first to create a free frame.

The implementation of FIFO has been explained with the help of an example.

Example 9.1: Consider a reference string 2,1,3,4,1,5,3,2,1,4,5. Suppose there are three free frames in memory initially. All frames are empty at the beginning. Calculate the total number of page faults and page replacements using FIFO.

Solution: The working of the algorithm is shown in Figure 9.4.

2, 1, 3, 4, 1, 5, 3, 2, 1, 4, 5

FIFO

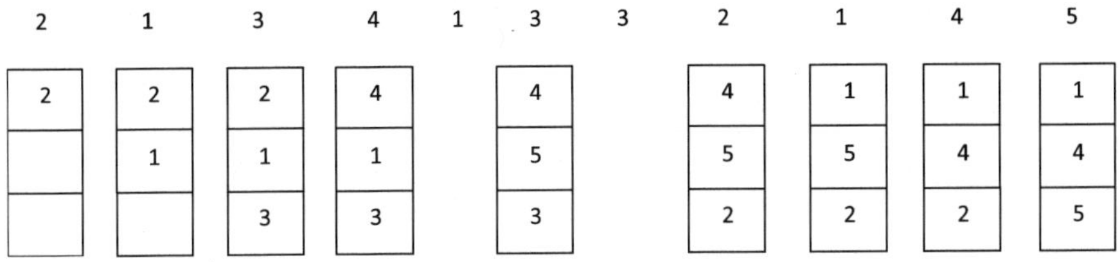

FIGURE 9.4. First In First Out algorithm

The total number of page faults and page replacements in the given example are as follows:

Page faults = 9
Page replacements = 6

Advantages of the First In First Out (FIFO) Algorithm

- It is a very simple algorithm.
- It is easy to implement.
- This algorithm can be coded easily.

Drawbacks of FIFO

- If the oldest page is used frequently, then performance of this algorithm decreases because the page swapped out will be required and it will result in the page fault.
- The FIFO algorithm suffers from Belady's Anomaly.

Belady's Anomaly

According to Belady's Anomaly, *the total number of page faults may increase with an increase in the number of free frames.* This can be proven with the help of an example.

Example 9.2: Consider Reference string 1, 2, 3, 4, 1, 2, 5, 1, 2, 3, 4, 5. Calculate the number of page faults using a FIFO algorithm having

a. 3 frames
b. 4 frames

Initially all frames are empty.

Solution:

a. The given reference string is implemented using three frames as shown in Figure 9.5.

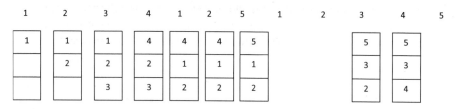

FIGURE 9.5. FIFO using 3 free frames

The total number of page faults and page replacements in the given example are as follows.

Page faults = 9
Page replacements = 6

b. The given reference string is implemented using four frames as shown in Figure 9.6.

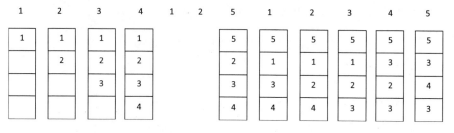

FIGURE 9.6. FIFO using 4 free frames

The total number of page faults and page replacements in the given example are as follows:

Page faults = 10
Page replacements = 7

As shown in the above example, in the FIFO algorithm, the number of page faults increases as the number of free frames are increased from 3 frames to 4 frames. This anomaly of result is called Belady's Anomaly.

⊙ Self-Quiz

Q9. Why is page replacement required?
Q10. What is Belady's Anomaly?
Q11. What is the working principle of the First In First Out algorithm?

Optimal Page Replacement

In this algorithm the page in the main memory which will be not be used for the longest time is swept out from the main memory to create space for the requested page. In this method to select the victim page, the future reference screen must be seen. The page which is not to be used in the near future will be removed from the main memory to create a free frame. *In optimal page replacement, remove the page that will not be used for the longest period of time.* The implementation of optimal page replacement has been explained with the help of an example.

Example 9.3: Consider a reference string 2, 1, 3, 4, 1, 5, 3, 2, 1, 4, 5. A reference string means that pages are referred from the main memory in this sequence. Suppose there are three free frames in memory initially.

Solution: The working of the algorithm is shown in Figure 9.7.

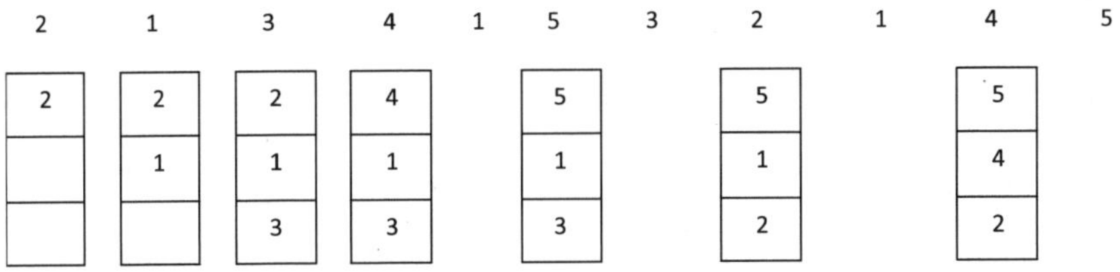

FIGURE 9.7. Optimal Page Replacement algorithm

The total number of page faults and page replacements in the given example are as follows:

Page faults = 7

Page replacements = 4

Advantages of the optimal page replacement algorithm are as follows:

- This algorithm produces a minimum number of page faults.
- It improves the system performance by reducing overhead for the handling of page faults.

Disadvantages of the optimal page replacement algorithm are as follows:

- This algorithm is difficult to implement.
- To implement this method the future reference should be known, which is not practical.

Least Recently Used Algorithm

In this algorithm the page that has not been used for the longest time is swapped out of the main memory to create space for the requested page. *Replace the page which has not been used for a very long time.*

Implementation of the LRU is done in many ways. One way is to implement it using an array. The array stores the information about the pages present in the main memory. The front end of the array stores the page which has been accessed recently. The rear end of the array stores the page that has not been accessed for the longest period of time. However, the page which is present in the main memory is accessed the information about the page in the array is shifted to the front end of the array. If the page fault occurs, the page indicated by the rear end of the array is swapped out of the main memory and the requested page is swapped in. The implementation of LRU has been explained with the help of an example.

Example 9.4: Consider a reference string 2, 1, 3, 4, 1, 5, 3, 2, 1, 4, 5. Reference string means that pages are referred from the main memory in this sequence. Suppose there are three free frames in memory initially.

Solution: The working of the algorithm is shown in Figure 9.8.

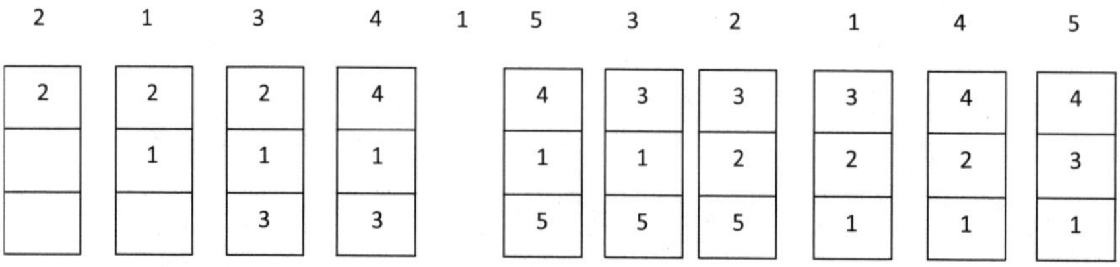

2	1	3	4	1	5	3	2	1	4	5
2	2	2	4	4	3	3	3	4	4	
	1	1	1	1	1	2	2	2	3	
		3	3	5	5	5	1	1	1	

FIGURE 9.8. Least Recently Used algorithm

The total number of page faults and page replacements in the given example are as follows:

Page faults = 9
Page replacements = 6

Advantages of the Least Recently Used Algorithm are as follows:

■ It is a very feasible method to implement.
■ It is not as simple as the FIFO algorithm, but it is not as complicated as optimal page replacement.

Disadvantages of the Least Recently Used Algorithm are as follows:

■ It requires additional data structure to maintain the array.
■ Extra hardware support is required.

⊙ Self-Quiz

Q12. What is the main drawback of the optimal page replacement algorithm?

Q13. What is the working principle of the LRU algorithm?

Second Chance Algorithm

■ This algorithm is a variation of the FIFO algorithm. In this method the page which is present for the longest time in the memory is given a second chance to remain present in the main memory. When that page is encountered a second time, it is taken out to create room for the page that has caused the page fault. *This algorithm is also known as the clock algorithm.*

■ This algorithm is implemented using the concept of a circular queue. Each cell in the circular queue contains two values, a page number and its corresponding reference bit. The value of the reference bit can be either 0 or 1. If the value of the bit of the page is 1, it means that page is encountered as the oldest page and allowed the second chance.

■ If the value of the reference bit is zero, this indicates that the oldest page has not been encountered yet. When the oldest page in the memory is found for the first time, the reference bit is set from 0 to 1. The next time that page is found, it is swept out of the main memory.

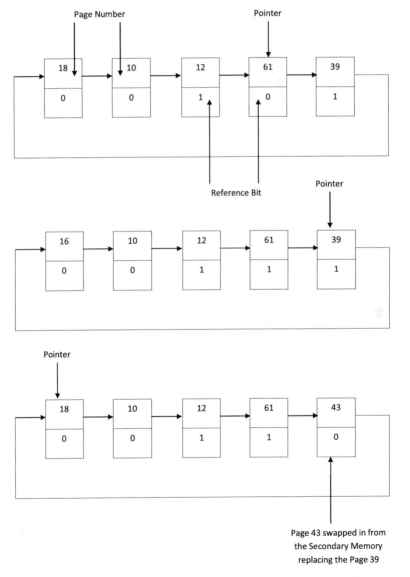

FIGURE 9.10. Implementation of the second chance algorithm

As shown in Figure 9.10, the pointer is directly at page 61. The page fault occurs. Page 61 has not been given the chance to remain in the memory, because its reference bit is 0. The reference bit of page 61 is made to be 1, and the pointer is moved to the next page which is 39.

The reference bit of page 39 is 1. This indicates that page 39 has previously been given the chance to remain loaded in the memory. This page is swept out of the memory to create space for the requested page 43.

Enhanced Second Chance Algorithm

This algorithm is a variation of the second chance algorithm. In this method, instead of using a reference bit, two bits are used. One bit stores the information about whether the page was recently used or not. If this bit is 0 it means the page has not been used recently. If the bit is 1 it indicates that the page has been used recently. *This is known as the usage bit.*

The second bit stores information about whether or not a page has been modified. If its value is zero it indicates that it has not been modified in the recent past. If the value is 1 then it indicates that the page has been modified. This bit is known as a *modification bit.*

One bit for each page is set by hardware whenever the page is modified. If a dirty page is replaced, it must be written to a disk before its page frame is reused. The bit used to indicate a modified page is called a dirty bit.

The usage bit and modification bit are represented as a pair.

■ The combination of [0, 0] indicates that the page has not been used or modified. This type of page is the best option of page replacement.
■ The combination of [0, 1] indicates that the page has not been used but has been modified. This type of page is the second best option of page replacement.
■ The combination of [1, 0] indicates that the page has been used but not modified. This type of page is the third best option of page replacement.
■ The combination of [1, 1] indicates that the page has been used as well as modified. This type of page is not a good option of page replacement.

Counting-Based Algorithms

Some page replacement utilizes the method of counting how many times a page has been used. Such an algorithm is known as a counting-based algorithm.

Examples of counting-based algorithms are the least frequently used algorithm and the most frequently used algorithm.

Least Frequently Used Algorithm (LFU): In the LFU algorithm the page that was referred to the lowest number of times from the time when the page is loaded in the main memory is replaced when the page fault occurs. The logic behind this algorithm is that some pages are accessed more frequently than others.

Most Frequently Used Algorithm (MFU): The most frequently used algorithm has reversed the idea of the LFU method. In the MFU the page that has been referred to the most is replaced. The idea behind the algorithm is that the page that has been referred to the lowest number of times may have been loaded recently and would be required soon.

ALLOCATION OF FRAMES

In multiprogramming environments, many processes are to be present in the main memory. Each process should get a reasonable amount of free memory space so as to have a minimum number of page faults. The number of frames in memory is limited. In this case with a limited number of frames and many processes, how would the operating system allocate the free frames to process pages? There should be some techniques in which the operating system allocates a minimum number of free frames to each process in the main memory, because each process needs a *minimum* number of pages. This is known as the allocation of frames. There are two major allocation schemes used by an operating system for the allocation of frames. They are as follows:

- Fixed allocation
- Priority allocation

Let us briefly discuss each one.

Fixed Allocation

There are two types of fixed allocation techniques.

- *Equal Allocation–* In the equal allocation technique, each process will get an equal number of frames. The process size will not affect the number of frames allotted to it. For example, if there are 10 processes in the system and 200 free frames in the main memory, according to equal allocation each process will get 20 frames each.
  The drawback of the equal allocation technique is it does not consider the size of the process in allocating the frames. Whether the process is small or large, it will get an equal number of frames. Hence the larger processes will face frequent page faults.
- *Proportional Allocation–* The drawback of the equal allocation technique is removed in this method. The idea behind this method is the *number of frames allotted to a process should be according to the size of the process.*
  If total number of available frames is "m," then ai frames are allotted to process p_i, where a_i is

$a_i = s_i / S \circ m$
Where $S = " s_i$

Example 9.5: Suppose there are two processes, P1 and P2, with sizes of 20 KB and 137 KB each. The number of free frames available is 72. Calculate the total number of frames allotted to each process using

  a. Equal allocation
  b. Proportional allocation

Solution:

  a. In equal allocation method process P1 will get 36 frames and process P2 will also get 36 frames. The requirement of P1 is only of 20 frames; hence, it is useless to allot 36 frames to P1. That is why the equal allocation method is not used.

  b. Applying the formula $a_i = s_i / S \,{}^\circ\, m$

    a1 = (20/157) ° 72
      = 9.1
      = 9 frames approximately
    Hence P1 will get 9 frames.

    a2 = (137/157) ° 72
      = 62.8
      = 63 frames approximately
    Hence P2 will get 63 frames.

Priority Allocation

- This method is a variation of the proportional method.
- In this method, the proportional allocation scheme is used, but by the prioritizing of the processes rather than by size.
- If process *Pi* generates a page fault, a frame from a process is selected for replacement with a lower priority number.

Global vs. Local Replacement

Global Replacement

- A frame from any process can be selected for replacement.
- A process selects a replacement frame from the set of all frames; one process can take a frame from another.
- Higher priority processes may get frames from lower priority processes.
- The total number of frames allocated to each process will change in this method as a process can get frames from any other process.

Local Replacement

- Each process selects from only its own set of allocated frames.
- A process cannot take frames from other processes.
- The total number of frames allocated to each process will not change in this method as a process cannot get frames from any other process.

Self-Quiz

Q14. Why is the proportional allocation method a better approach as compared to equal allocation?
Q15. What is the priority frame allocation method?
Q16. What is the main advantage of the priority frame allocation method?

THRASHING

If a page fault occurs every time page is referred, then the system is said to be thrashing. Thrashing is directly related to the rate of page fault occurrence. The higher the page fault rate, the more the process will be in a thrashing state. Thrashing degrades the performance of the system. When thrashing occurs, the system performance downgrades. This happens because the processor spends more time in handling page faults rather than actual processing of instructions present in the pages. Thrashing has a negative effect on the system performance.

When the page fault rate increases, it needs to be handled by page fault handling strategies. The steps followed in handling page faults will need processing from the paging device. Hence, the queue at the paging device increases. This will result in increased service time for a page fault. In other words, when the page fault rate is high, then time taken for handling page faults also increases. Hence the steps/transactions in the system will wait for the paging device. This will decrease CPU utilization and system throughput and increase system response time, resulting in below optimal performance of a system.

As the degree of multiprogramming of the system increases beyond a certain level, thrashing occurs. Hence thrashing is a threat when the degree of multiprogramming of the system increases. This situation is shown with the help of Figure 9.11.

The graph shown in Figure 9.11 indicates that:

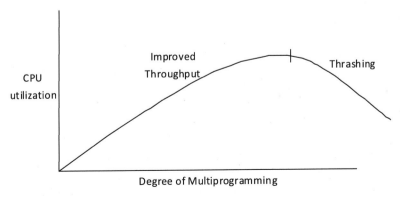

FIGURE 9.11. Thrashing with increase in degree of multiprogramming

- Initially when there is an increase in the degree of multiprogramming, there is also improved throughput.
- Increased throughput is desired for optimal system performance.
- CPU utilization reaches a maximum point as the degree of multiprogramming increases.
- After reaching that point there is a decline in throughput.
- From that point a sharp decline in throughput occurs and thrashing sets in.
- Hence thrashing occurs in the overextended system (a large number of processes in the main memory).

METHODS TO PREVENT THRASHING

As explained in Section 9.6, thrashing has a negative effect on the system performance. Hence it needs to be prevented.

Various strategies can be used to eliminate the occurrence of thrashing. The various methods are:

- By controlling the load on the system
- Proper selection of a replacement policy to implement virtual memory
- Working set strategy
- Concept of pre-paging
- Page fault frequency

By Controlling the Load on the System

As shown in Figure 9.11 the load increase in the main memory results in thrashing. Hence if a limited number of processes is kept in the main memory, then this may help in preventing thrashing. That is why multiprogramming should only be increased up to a certain level beyond which it will not be desirable. The aim of increasing a degree of multiprogramming is to get improved throughput. If an increase in the degree of multiprogramming results in thrashing and decline in throughput, then it is preferable to have a lower degree of multiprogramming.

Proper Selection of a Page Replacement Policy

There are various page replacement policies in virtual memory. The proper selection of a replacement policy to implement virtual memory can play an important role in the prevention of thrashing. Local page replacement is considered to be a better approach than global page replacement. It limits the effect of thrashing. In *Local replacement* each process selects from only its own set of allocated frames. A process cannot take frames from other processes. The total number of frames allocated to each process will not change in this method, as processes cannot get frames from any other process. Whereas in *Global replacement*, a frame from any process can be selected for replacement. A process selects a replacement frame from the set of all frames; one process can take a frame from another. A process of higher priority may get frames from low priority processes. The total number of frames allocated to each process will change in this method as the process can get frames from any other process. *The replacement policy which uses the global replacement method is more likely to cause thrashing.*

Working Set Strategy

The set of pages that a process is currently using is called the working set of the process. If the working set of a process is known, then pre-paging can be done, which will help in reducing page faults. This strategy will directly help in preventing thrashing in a system. *Working sets* is a conceptual model proposed by Peter Denning to prevent thrashing.

■ If the sum of all working sets of all processes exceeds the size of memory, then stop running some of the processes for a while.

■ Divide processes into two groups: active and inactive:

- When a process is active its entire working set must always be in memory: never execute a thread whose working set is not resident.
- When a process becomes inactive, its working set can migrate to secondary memory.
- Threads from inactive processes are never scheduled for execution.
- The collection of active processes is called the *balance set*.
- The system must have a mechanism for gradually moving processes into and out of the balance set.
- As working sets change, the balance set must be adjusted.

How to Compute Working Sets

■ Denning proposed a working set parameter T. In the proposed model all pages which are referenced in the last T seconds will be present in the working set.

■ A clock algorithm can be used and extended to keep an *idle time* for each page.

■ Pages with idle times less than T are in the working set.

The Concept of Pre-Paging

The set of pages that a process is currently using is called a working set of the process. If the working set of a process is known, then pre-paging can be done, which will help in reducing page faults. The pages which will be required in the near future can be fetched in advance and kept in the main memory. This is known as pre-paging. This strategy will directly help in preventing thrashing in a system.

Page Fault Frequency

In this method the frequency of page faults can be observed. If a particular process is not facing page faults, then it indicates that the process has a sufficient number of pages allotted to it. On the other side if another process is facing a lot of page faults, then it indicates that the process has an insufficient number of pages allotted to it. In such situations, pages from processes with less faults can be taken and allotted to the process which is facing frequent page faults. Hence by observing the frequency of page faults, the occurrence of thrashing can be controlled.

Although there are many strategies to limit thrashing, it cannot be prevented fully. If a single process is too large for memory, then the OS cannot do much in preventing thrashing. That process will simply thrash.

In practice, today's operating systems do not worry much about thrashing. In personal computers, users can notice thrashing and handle it themselves by just buying more memory. They can manage balance sets manually. In fact thrashing is a bigger issue for timesharing machines with many users. In timesharing systems, the users will be reluctant to stop their programs so that others can progress. Also such a system had to handle thrashing automatically. In fact technology advancements have made changes to operate machines in a range where memory is even slightly overcommitted because it is better to just buy more memory.

DEMAND SEGMENTATION

Demand segmentation is similar to the concept of demand paging, but with a few changes. In demand segmentation memory is allocated to the processes in the form of segments. Segment descriptors are maintained in memory (like page tables) to keep track of segments. Segment descriptors have details of the segments in which processes are stored. They contain a field called the valid/invalid bit. Its value can be 0/1. If the value of this entry is 1, it indicates that the segment is present in the main memory; otherwise, valid segments are not present in the main memory and a segment fault arises. As segments are of different sizes (unlike pages), base and limit are stored in the segment table.

There are some differences between demand paging and demand segmentation, which are as follows:

- In demand paging, the available memory space is divided into equal size pages, whereas in demand segmentation, memory is divided in various-sized segments.
- The demand paging method is mainly used to implement virtual memory, because segment replacement algorithms are more complex than page replacement algorithms.
- Segment replacement algorithms are more complex than page replacement algorithms (will be explained in section 9.4), because segments are of different (variable) sizes.
- The programmer is aware of demand paging, but they are not aware of the presence of demand segmentation.
- There is only one address space in demand paging, yet there may be many address spaces in demand segmentation.

- Internal fragmentation is present in demand paging, but it is removed in demand segmentation.
- External fragmentation is absent in demand paging, but it is present in demand segmentation.
- Despite its complexity, demand segmentation is used in some operating systems. The IBM OS/2 operating system uses demand segmentation.

⊙ Self-Quiz

Q17. Define thrashing.?
Q18. What is the main cause of thrashing?
Q19. How can thrashing be detected?
Q20. How can thrashing be removed?

SUMMARY

Virtual memory is considered as an illusion that a computer system possesses more memory than it actually has. This concept makes a process independent of the size of the real main memory. It allows many processes to be present in the main memory for executing without being restricted by its size. Various concepts related to virtual memory have been explained in the chapter.

POINTS TO REMEMBER

- A technique that allows the execution of processes that may not be completely in the main memory is called virtual memory.
- To allow the programmer to make programs larger than the main memory size, the concept of virtual memory has been used.
- Virtual memory is implemented by demand paging and demand segmentation.
- In demand paging, the available memory space is divided into equal-size pages.
- In demand segmentation memory is divided in various-sized segments.
- The paging technique in which pages are loaded in the main memory according to demand or need is called demand paging.
- When a desired page is not present in the main memory, then such a condition is known as a page fault.
- For better performance of demand paging, the effective access time should be as small as possible.
- The smaller the value of EAT, the better the performance of demand paging.
- Selecting some page in memory that may not be in use and swapping it out creates a free frame in memory, and placing the desired page in that newly created free frame is called page replacement.
- There are many page algorithms, for example, first in first out, optimal page replacement, and least recently used.

- In the FIFO algorithm, swap out the page which was entered into memory first to create a free frame.
- The FIFO algorithm suffers from Belady's Anomaly.
- According to Belady's Anomaly, the total number of page faults may increase with an increase in number of free frames.
- In optimal page replacement, remove the page that will not be used for the longest period of time.
- In LRU replace the page which has not been used for a very long time.
- In the second chance algorithm, the page which is present for the longest time in memory is the given the second chance to remain present in the main memory. When that page is encountered for a second time it is taken out to create room for the page that has caused the page fault.
- The second chance algorithm is also known as the clock algorithm.
- To allocate a minimum number of free frames to each process in the main memory because each process needs a minimum number of pages is known as the allocation of frames. When the processor spends more time in handling page faults rather than the actual processing of instructions present in the pages, then this situation is known as thrashing.

REVIEW QUESTIONS

Q1. Discuss the following in detail.
   a. Demand paging
   b. Thrashing
Q2. What do you understand about page replacement? Name the algorithms available for page replacement.
Q3. How many page faults occur in the optimal page replacement algorithm with the following reference string for four page frames?
   1,2,3,4,5,3,4,1,6,7,8,7,8,9,7,8,9,5,4,5,4,2
Q4. Consider the following page reference string 4,3,2,1,4,3,5,4,3,2,1,5. How many page faults would occur for the following page replacement algorithms assuming three frames? Initially all frames are empty.
   a. FIFO
   b. LRU
   c. Optimal
Q5. What is thrashing? How is it overcome?
Q6. Define virtual memory concepts and also discuss page replacement algorithms in brief.
Q7. When does page fault occur? Describe in detail the actions taken by the operating system when a page fault occurs.
Q8. What is the cause of thrashing? How does a system detect thrashing? Once it detects thrashing, what can the system do to eliminate this problem?
Q9. Consider a demand paged system. Page tables are held in registers. It takes 8 milliseconds to handle a page fault if an empty page is available

or if the replaced page is not modified. The memory access time is 100 nanoseconds. Assume that the page to be replaced is modified 70 percent of time. What is the minimum acceptable page fault for an effective access time of no more than 200 nanoseconds?

Q10. Explain advantages and disadvantages of thrashing.

Q11. Consider the following pages in a reference string.

1,2,0,3,5,1,5,7,2,0,3,5,4,1,2,5,3,7

Implement FIFO, LRU, and optimal page replacement algorithms and calculate the total number of page faults.

Q12. Consider the following pages in a reference string.

1,2,3,4,2,1,5,6,2,1,2,3,7,6,3,2,1,2,3,6

Implement FIFO, LRU, and optimal page replacement algorithms and calculate the total number of page faults using three and four frames.

10

FILE SYSTEMS

Purpose of This Chapter

The purpose of this chapter is to explain the function of file systems, to describe the interfaces to file systems, to discuss file-system design tradeoffs, including access methods and directory structures, to describe the allocation methods, and to explore file-system protection.

INTRODUCTION

In an operating system the file system is the most noticeable aspect. The mechanism provided by the file system is to store online both the data and the programs of the operating system and to access those data and programs. The file system contains two different parts, a collection of files (each file storing related data) and a directory structure (that organizes and provides information about all the files in the system). This chapter emphasizes the various aspects of files and the main directory structures, and describes the methods to allocate disk space, recover freed space, track the locations of data, and interface other parts of the operating system to secondary storage. Last, it will discuss the ways of handling file protection. File protection is required when there are multiple users and controlling is required to determine who may access files and how the files may be accessed.

CONCEPT OF FILE

A disk can be considered as a linear sequence of fixed-size blocks. To explain the file (a logical storage unit), the physical properties are abstracted by the operating system from its storage devices. A file is a named collection of related information. All this information is recorded on secondary storage as a sequence of bytes. The file can be viewed as logical and physical. The logical or programmer's view depicts how the files can be used and what properties they have (i.e., as the users can see it) whereas the physical or operating system's view represents how these files actually exist in secondary storage.

A file represents the programs (programs include source and object forms) and the data (a data file can be numeric, alphanumeric, alphabetic, or binary). A file is a sequence of bits, bytes, lines, or records. All the information that is stored in a file is defined by the creator of that file. A file can store many different types of information, such as source programs, object programs, executable programs, numeric data, text, payroll records, graphic images, sound recordings, and so on.

We know that a file can store different information. Therefore the structure of a file may also vary depending on its type. The file is said to be a text file when the sequence of characters is organized into lines or pages. When there is a sequence of subroutines and functions in a file, and these subroutines and functions are further organized as declarations followed by executable statements, then the file is said to be a source file. A file that contains a sequence of bytes organized into blocks explicable by the system's linker is called an object file. A file that holds a series of code sections that the loader can bring into memory and execute is known as an executable file.

Attributes of a file: Although the attributes of a file may vary depending on the operating system, usually the file consists of the following attributes:

- **Name:** The file is named so that it will be convenient for users. The name of the file is the only information that is in the human readable format.
- **Identifier:** For any file identifier, the name cannot be read by the users, as it in the non-human readable format. The identifier is a unique tag (usually a number) that identifies the file within the file system.
- **Type:** This information is required by those systems that support different types of files.
- **Location:** This information tells the location of a file stored on the device. Therefore it is a pointer to a device and to the location of the file on that device.
- **Size:** This field gives the current size of the file or the maximum allowed size for any file in bytes, words, or blocks.
- **Protection:** This field provides the access-control information, that is, it determines that who can read, write, or execute a file.
- **Time, date, and user identification:** This information keeps track of when and who created, last modified, or last accessed the file. All this information is useful for protection, security, and usage monitoring.

The information about all the files is kept in the directory structure. This directory structure also resides on the secondary storage. A directory contains the name of the file and its unique identifier. Other attributes of the file will be located by the identifier.

Operations on file: A file is an abstract data type. The proper definition of a file can be provided on the basis of the operations that can be performed on a file. The six basic file operations are create, read, write, reposition, delete, and truncate. To perform all these operations, the operating system can provide system calls.

■ **Create:** For creating a file, two steps are required: (a) Space for the file must be found in the file system, and (b) An entry regarding the new file must be made in the directory.

■ **Write:** For writing a file, a system call is made that specifies both the name of the file and the information to be written to the file. A write pointer must be kept by the system at the location where the next write is to take place. Whenever a write takes place, the write pointer must be updated.

■ **Read:** To perform the read operation on a file, a system call must be used, which specifies the name of the file and the location in memory (i.e., where the next block of the file should be put). Also, a read pointer is required that is kept at that location in the file where the next read is to take place. The read pointer is updated after the read has taken place. Typically, a process can be either reading from or writing to a file, and therefore the current operation location can be kept as a per-process current file-position pointer. This same pointer can be used by both the read and write operations, which saves space and reduces system complexity.

■ **Reposition:** The directory is searched for the appropriate entry, and the current-file-position pointer is repositioned to a given value. No Input/ Output is involved in repositioning within a file. This file operation is also known as a file seek.

■ **Delete:** For deleting a file, the directory is searched for the named file. After finding the associated directory entry, all the file space is released (so that this space can be reused by other files) and the directory entry is erased.

■ **Truncate:** There may be a case when the user may want to erase the contents of a file but keep its attributes. Truncating a file allows all attributes to remain unchanged except the length of file (the length is reset to zero and its file space is released).

These six operations are the basic required file operations. Other than these basic operations, some common operations are **appending** new information at the end of an existing file and **renaming** an existing file. All these operations can then be combined to perform other file operations such as to create a **copy** of a file or to **copy** the file to another I/O. Some operations are also required that allow a user to get and set the various attributes of a file.

The OS maintains a small table, known as the open-file table, which contains the information about all open files. Whenever a file operation is requested, the file is specified in this table through an index number. Therefore, overhead of searching is not required. When the file is not being used, it gets closed by the process, and its entry from the open-file table is removed by the operating system.

Most of the systems require that the programmer open a file explicitly using the open() system call before that file can be used. The open() operation takes a file name and starts searching the directory, by copying the directory entry to the open-file table. Access-mode information (such as create, read-only, read-write, append-only, and so on) can also be accepted by this system call. The file will be opened for the process, if the request mode is allowed. Usually the open() system call returns a pointer to the entry in the open-file table. This pointer is used in all I/O operations, which avoids further searching and simplifies the system-call interface.

To implement the open() and close() operations in an environment where several processes may open the file at the same time is complicated. This condition arises in those systems where at the same time several different applications open the same file. Usually two levels of internal tables are used by an operating system, that is, *a per-process table* and *a system-wide table*. The per-process table keeps track of all the files that a process has open. For example, the current file-position pointer for each file is found here. It can also include the access rights to the file and the accounting information. Each entry in the per-process table points to a system-wide table in sequence. The system-wide table contains information that is process independent, such as the file location on disk, access dates, and file size. Once a process opens the file, the system-wide table includes an entry for that file.

Usually, the open-file table also has an open count which is associated with each file to specify how many processes have the file open. Each close() system call decreases this open count, and when the open count becomes zero, it means that the file is no longer in use, and the entry is removed from the open-file table. The following is some information associated with an open file: file pointer, file-open count, disk location of the file, and access rights.

TABLE 10.1. Common File Types

file type	usual extension	function
executable	exe, com, bin or none	ready-to-run machine-language program
object	obj, o	compiled, machine language, not linked
source code	c, cc, java, pas, asm, a	source code in various languages
batch	bat, sh	commands to the command interpreter
text	txt, doc	textual data, documents
word processor	wp, tex, rtf, doc	various word-processor formats
library	lib, a, so, dll	libraries of routines for programmers
print or view	ps, pdf, jpg	ASCII or binary file in a format for printing or viewing
Archive	arc, zip, tar	related files grouped into one file, sometimes compressed, for archiving or storage
multimedia	mpeg, mov, rm, mp3, avi	binary file containing audio or A/V information

Some of the operating systems provide facilities for locking an open file or sections of a file. File locks allow one process to lock the file and prevent its access from other processes. File locks are useful for the files that are shared by several processes.

Types of File: To implement a file type, the most common technique is to include the type as a part of the file name. The name of a file is split into two parts, a name and an extension, both separated by a period character as shown in Figure 10.1. Therefore, the user and the operating system can tell the type of the file from its name only. For example, *schedule.doc, Server.java,* and *ReadThread.c.* The extension of the file is used by the system to indicate the file type and the type of operations that can take place on that file. For example, only the files with a *.com, .exe,* or *.bat* extension can be executed. The *.com* and *.exe* files are the two forms of binary executable files, and a *.bat* file is a batch file that contains (in ASCII or American Standard Code for Implementation Interchange format) commands to the operating system. Microsoft Disk Operating System or MS-DOS recognizes only a few extensions.

File extensions are also used by the application programs to specify the type of the file in which they are interested. For example, the assemblers expect source files to have an *.asm* extension, and the Microsoft Office Word processor expects files to end with a *.doc* extension. As the operating system does not support these extensions, they can be thought of as "hints" to the applications that operate on them.

Many of the file systems support names as long as 255 characters. Some file systems differentiate between upper and lower case letters (i.e., case sensitive and case insensitive). Both Windows 95 and Windows 98 use the Microsoft Disk Operating System or MS-DOS file system and therefore inherit many of its properties, such as how the file names are constructed. Also, Windows NT and Windows 2000 support the MS-DOS file system, so these two also inherit its properties. However, these two systems also have a Native File System or NTFS that has different properties (such as file names in Unicode).

Consider the Mac OS X OS, where each file has a type, such as TEXT (for text file) or APPL (for application). Here each file also has the attribute of a creator, which contains the name of the program that created it. The operating system sets this attribute during the *create()* system call, and therefore its use is enforced and supported by the system.

The UNIX system uses a basic *magic number* that is stored at the beginning of some files to indicate roughly the type of file, such as executable program, batch file, PostScript file, and so on. As all the files do not contain the magic number, the features of a system cannot be based only upon this information. Also, UNIX does not keep track of the name of the creating program. File-name-extension hints are not allowed in UNIX, but these extensions are neither enforced nor depended on by the OS; typically, they are destined to help the users to determine the type of contents of the file. Depending on the application the extensions can be used or ignored, but this is totally up to the application's programmer.

Internal File Structure: To locate an offset within a file can be complicated for the operating system. Usually disk systems have a well-defined block size, which is determined by the size of a sector. The entire disk I/O is performed in units of one block (physical record), and all blocks are of the same size. It is improbable that the size of the physical record matches exactly the length of the desired logical record. One common solution to this problem is to pack a number of logical records into physical blocks.

For example, the UNIX operating systems defines all files to be the streams of bytes, where each byte is individually addressable by its offset from the beginning or end of the file. In this case, the size of the logical record is 1 byte. The file system packs and unpacks bytes automatically into physical disk blocks (512 bytes per block) as essential.

The file may be considered as a sequence of blocks, and all the basic I/O functions operate in terms of the blocks.

Disk space is always allocated in blocks; therefore, some portion of the last block of each file is wasted. For example, if each block were of 512 bytes, then a file of 1,949 bytes would be allocated four blocks (i.e., 2,048 bytes) and the last 99 bytes would be wasted. To allocate the complete file into blocks rather than bytes is called the internal fragmentation. All the system files suffer from *internal fragmentation*, that is, the larger the block size, the greater the internal fragmentation.

⊙ Self-Quiz

Q1. Explain file attributes.
Q2. What are the operations that can be performed on a file? Explain.
Q3. Write a short note on the types of files.

ACCESS METHODS

The information that is stored in a file can be accessed in various ways. Some systems can access the files using only one access method, whereas other systems (such as IBM or International Business Machines), support various access methods for accessing the file. Choosing the correct access method for any particular application is a major designing problem.

Sequential Access

This is the simplest access method. In this method, processing of the information that is stored in a file must be done in order, that is, one record after the other. This is the most common method for accessing the file. For example, the editors and the compilers of a computer usually access the files in the following manner:

On a file, the reads and writes operations are used in bulk. A read operation, that is, "read next," reads the next portion of the file and updates the file pointer automatically, to which the I/O location would be tracked. In the same way, the write operation, that is, "write next," appends to the end of the file and advances to the end of the newly written material. Such a file can be reset to the beginning, and on some systems, a program may be able to skip forward or backward n records for some integer n. This method is called sequential access to the file and is based on a tape model of a file.

Direct Access

Another method for accessing a file is direct access, also known as the relative access method. A file is a collection of fixed length logical records, which allow programs to read and write records quickly in a random order. The direct access method is based on a disk model of a file, because disks allow random access to any block of a file. For direct access, the file can be viewed as a numbered sequence of blocks or records. For example, after reading block 17, the program could read block 58 next, and then write block 9. In a direct access method, there is no restriction on the order of reading or writing.

Direct access files can be used when large amounts of information require immediate access. An example is in databases, where when a query arrives, the block that contains the answer is computed first and then the read operation takes place for that block directly to provide the desired information.

In the case of the direct access method, the file operations must be modified for the purpose of including the block number as a parameter. Therefore, we have "read n" rather than "read next," and "write n" rather than "write next," where "n" is the block number. An alternative approach is to keep the "read next" and "write next" (as it is done in sequential access) and add an operation to "position file to n," where n is the block number. Then, to affect a "read n," first the "position to n" and then the "read next" commands are issued.

All the operating systems do not support both the sequential and the direct access for files. Some systems allow only sequential file access, whereas others allow only direct access. Some of the systems require that when a file is created it must be defined as sequential or direct. This type file can be accessed only by that method that was declared at file creation time.

Other Access Methods

On the top of the direct access method, other access methods can be built. In these additional methods an index for a file is created, which contains pointers to various blocks. Therefore, to find an entry in the file, first the index is searched and then the pointer is used to access the file directly for the purpose of finding the desired entry.

In the case of large file, the index for that file also becomes too large to store in memory. This problem can be solved by creating an index for the index file. In this case the primary index file would contain pointers that point to the secondary index files, and the secondary index files would contain pointers that point to the actual data items.

For example, a small master index is used by IBM's Indexed Sequential Access Method (ISAM) that points to disk blocks of a secondary index, and the secondary index blocks point to the actual file blocks. The file is kept sorted on a defined key. To search for a particular item, first a binary search should be applied on the master index, through which the block number of the secondary index is found. This block is read in, and again a binary search is applied to find the block containing the desired record. Finally, this block is searched for in a sequential manner. Though this method, any record can be located from its key by most direct access reads.

⊙ Self-Quiz

Q4. **What do you mean by access?**
Q5. **Explain the sequential access method.**
Q6. **What are the advantages of the direct access method over the sequential access method? Explain.**

DIRECTORY STRUCTURE

A system may have zero or more file systems, and the file systems can be of various types. Therefore, the file systems of computers can be widespread. Some of the systems may store millions of files on Terabytes (TB) of disk. To manage all these data, it is necessary to organize them. Using a directory is required for organizing those data.

Storage Structure

For a file system, a disk may be used completely. Therefore, for storing multiple file systems a whole disk or a part of a disk (where the other part can be used for other things, such as swap space or unformatted disk space) is required. On each part of the disk, a file system can be created. The parts of a disk are known as partitions, slices, or minidisks (according to IBM). The part of the disk that holds the file system is referred to as a volume, and a volume can be thought of as a virtual disk. Multiple operating systems can also be stored on a volume that allows a system to boot and run more than one.

If a volume contains a file system, then it must also contain information about the files in the system. All this information is kept in entries in a directory or volume table of contents. The directory records information such as name, type, address, current length, maximum length, date last accessed, date last updated, owner ID, and protection information for all files on that volume. A typical file-system organization is shown in Figure 10.1.

Directory Overview

The directory can be viewed as a symbol table that translates the file names into their directory entries. The following are the operations that have to be performed on a directory when considering a particular directory structure:

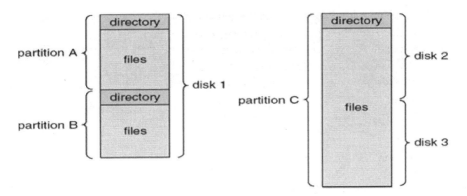

FIGURE 10.1. A typical file system organization

- **Search for a file:** Need to find a particular file or find all the files whose names match a particular pattern.
- **Create a file:** Need to create a new file and to add its entry to the directory.
- **Delete a file:** Need to remove a file from the directory, when that file is no longer required.
- **List a directory:** Need to list both the files in the directory and the directory contents for each file.
- **Rename a file:** Need to rename a file when the contents or use of the file changes. Renaming a file may involve changing the position of the file entry in the directory structure.
- **Traverse the file system:** Need to access every directory and every file within a directory structure. Therefore, for reliability, the directory needs a logical structure.

The most common schemes for defining the logical structure of a directory would be single-level directory, two-level directory, tree-structured directories, acyclic-graph directory, and general-level directory.

Single-Level Directory

The single-level directory is the simplest method for defining the logical structure of a directory. In the single-level directory structure all the files are contained in the same directory as shown in Figure 10.2. The single-level directory is easy to understand and to support.

This type of structure was commonly used in early computers (such as super computers), because there was only one user. A single-level directory does not work properly when the number of files on the system increases or when the system has more than one user. Because in this directory structure all files are in the same directory, their names must be unique. If two users name their data file *create*, then the unique-name rule is violated.

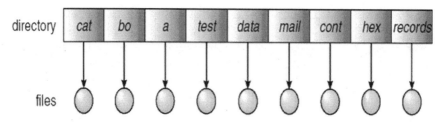

FIGURE 10.2. Single-level directory

Two-Level Directory

The problems of a single level directory can be resolved by creating a separate directory for each user. This standard solution is termed as a two-level directory structure, where each user has a separate *User File Directory (UFD)*. The structure of all the UFDs is similar, but each UFD contains the files of single user only.

When a user starts a job on the system, the *Master File Directory (MFD)* of that system is searched. The index for this MFD will be either the user name or the account number, and each entry of the MFD points to the UFD for that user.

Whenever a user wants to refer to a particular file (for the purpose of creation, deletion, alteration, etc.) then only that user's own UFD is searched to find that file. The two-level directory structure solves the name-collision problem, but it still has a drawback in that it isolates one user from the other. Isolating the users could be an advantage when the users are not dependent on each other, but is considered as a disadvantage when the users want to share files.

A two-level directory structure can be thought of as a tree, or an inverted tree, or a tree of height 2. The root of the tree is the MFD, the direct descendants of the root (MFD) are the UFDs, and the files are the direct descendants of these UFDs.

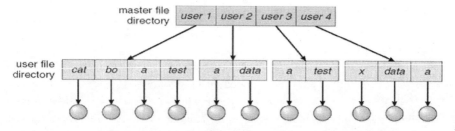

FIGURE 10.3. Two-level directory Structure

These files are the leaves of the tree. The path from the root (MFD) to the leaf (specified file) in the tree can be defined by specifying the name of the user and the name of the file. To specify the volume of a file, additional syntax is required. For example, in MS-DOS the volume of a file is specified by a letter followed by a colon. Hence, the specification of a file might be: *C:\user\create.*

Tree-Structured Directories

We have seen in a previous section how a two-level directory can be viewed as a two-level tree. To extend a directory structure to a tree structure of arbitrary height as shown in Figure 10.4 is a natural generalization. According to this generalization users are allowed to create their own subdirectories and organize their files consequently. The tree-structured directory is the most common directory structure.

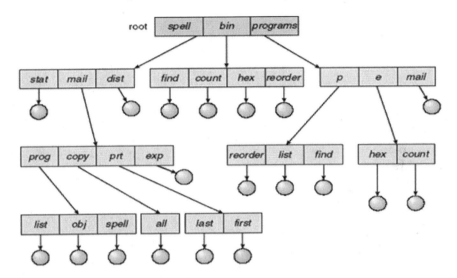

FIGURE 10.4. Tree-Structured directory

In this type of directory structure, the tree has a root directory, and all the files in the system have a unique path name. A directory is just another file, but is treated in a special way. The internal formats of all the directories are similar. One bit in each directory entry defines the entry as a file (0) or as a subdirectory (1).

The path name can be of two types, that is, *absolute* and *relative*. An absolute path name begins from the root and follows a path down to the specified file, giving the directory names on the path. The relative path name defines a path from the current directory.

With a tree-structured directory system, users can also be allowed to access the files of other users. For example, user Y can access a file of user X by specifying its path names. User Y can specify either an absolute or a relative path name. On the other hand, user Y can change the current directory to be user X's directory and access the file by its file names.

Acyclic-Graph Directories

The natural generalization of the tree-structured directory is the acyclic graph. In the acyclic graph, the common subdirectory should be shared. At certain times, a shared directory or file will exist in the file system in two (or more) places.

Sharing of files or directories is not allowed in a tree structure. An *acyclic graph* (i.e., a graph that does not have any cycle), allows the directories to share subdirectories and files as shown in Figure 10.5. The same file or subdirectory may be present in two different directories.

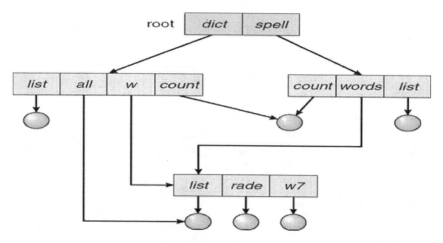

FIGURE 10.5. Acyclic-graph directory Structure

The shared file (or directory) and the two copies of the file are not the same thing. The two copies of the file imply that each programmer can view the copy rather than the original file, but if any one programmer changes the file, then those changes will not appear in the other's copy. Whereas in a shared file there exists only one actual file, so if any changes are made by one person on the file, then the changes will be immediately visible to the other.

There are various ways available to implement the shared file and subdirectories. As demonstrated by many of the UNIX systems, the most common way to implement the shared files and subdirectories is to create a new directory entry called a *link*. A link is a pointer to another file or subdirectory. The directory is searched when a reference to a file is made. If the entry of the directory is marked as a link, then the name of the real file is included in the link information. The link can be resolved by using that path name to locate the real file. Links are easily identified by their format in the directory entry and are effectively named indirect pointers.

Another common method to implement the shared files is to simply duplicate all the information about them in both the sharing directories. Therefore, both the entries are identical and equal. It is clear that a link is different from the original directory entry; hence, the two are not equal. When a file is modified, maintaining consistency is the major problem with duplicate directory entries.

There are several problems with the acyclic-graph directory structure. The first problem is that in this directory structure a file has multiple absolute path names, so distinct file names may refer to the same file. Another problem is regarding deletion, that is, when to deallocate or to reuse the space that is allocated to the shared file. One possible solution is to remove the file whenever anyone deletes it, but the removal of the file will leave dangling pointers, which will be pointing to the now non-existent file. The worst case occurs if the dangling pointers point to the middle of other files, when the remaining file pointers contain the addresses of actual disk and accordingly the space is reused for other files.

The use of symbolic links to implement sharing in the system is easy to handle. By deleting the link, only the link will be removed and the original file will not be affected at all. But if the entry of the file is deleted, then the allocated space for the file will be deallocated, and this will leave the dangling links. These links cannot be considered until an attempt is made to use them. And when these links are used, they point to the name of the file that does not exist and fail to resolve the link. This type of access is treated in a similar way as with any other illegal file name. In UNIX, symbolic links are left when a file is deleted, and it depends on the user to realize that the original file is deleted or has been replaced. Microsoft Windows also uses the same approach.

The other approach for deletion is to save the file until all the references to it are deleted. For implementing this approach, a mechanism is required for determining that the last reference to the file has been deleted. This approach is not very feasible because of the variability and the large size of the file-reference list.

However, it is not required to keep the entire list. The only requirement is to keep count of the number of references. The reference count is incremented when a new link or directory entry is added and is decremented when a link or directory entry is deleted. When the reference count is 0, it means that there is no remaining reference to it and the file can be deleted.

This approach is used by the UNIX operating system for non-symbolic links (or *hard links*), where a reference count is kept in the file information block (or *inode*). By effectively eliminating multiple references to directories, an acyclic-graph structure is maintained.

To avoid the problems discussed above, some systems do not support the concept of shared directories or links. For example, in MS-DOS, the directory structure is a tree structure rather than an acyclic-graph directory structure.

General Graph Directory

An acyclic-graph structure ensures that there are no cycles, which results in a major problem. Consider a two-level directory and allow the users to create subdirectories; this will result in a tree-structured directory. It is very simple to preserve the tree structure if the new files and directories are added to an existing tree structure. However, when the new links are added to an existing tree-structured directory, then the tree structure is destroyed, and it results in a simple graph structure as shown in Figure 10.6.

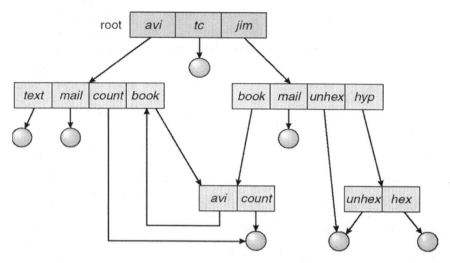

FIGURE 10.6. General Graph Directory

⊙ Self-Quiz

Q7. What do you understand by directory structure?

Q8. Explain the single-level directory structure.

Q9. What are the advantages and disadvantages of two-level directory structure?

Q10. Write short notes on:

    a. Tree-structured directory

    b. Acyclic-graph directory

FILE-SYSTEM STRUCTURE

Secondary storage is provided by disks, where a file system is maintained. Disks have two main characteristics that make them a suitable medium for storing multiple files:

1. A disk can be rewritten in place; it is possible to read a block from the disk, modify the block, and write it back into the same place.
2. A disk can access directly any given block of information it contains.

Therefore, it is easy to access any file either sequentially or randomly, and switching from one file to another requires only moving the read-write heads and waiting for the disk to rotate.

I/O transfer between memory and disk are performed in units of *blocks*, rather than transferring a byte at a time, to improve the I/O efficiency. Each block contains one or more sectors. Sectors vary from 32 bytes to 4,096 bytes (usually they are 512 bytes), depending on the disk drive.

For efficiently and conveniently accessing the disk, the operating system imposes one or more file systems to allow the data to be stored, located, and retrieved easily. For a file system the two design problems are:

1. To define how the file system should look to the user. This task involves defining a file and its attributes, the operations allowed on a file, and the directory structure for organizing the files.
2. Creating algorithms and data structures to map the logical file system onto the physical secondary-storage devices.

Generally, the file system is composed of many different levels. The structure shown in Figure 10.7 is an example of a layered design, where each level uses the features of lower levels to create new features for use by higher levels.

The *I/O control* consists of device drivers and interrupts handlers to transfer information between the main memory and the disk system. A device driver can be thought of as a translator, which takes high-level commands as input and gives output as low level, hardware-specific instructions. These instructions are used by the hardware controller, which is an interface between the I/O devices to the rest of the system. The device driver usually writes specific bit patterns to special locations in the I/O controller's memory to tell the controller which device location to act on and what actions to take.

The only requirement of the *basic file system* is to issue generic commands to the appropriate device driver for the purpose of reading and writing physical blocks on the disk. Each of the physical blocks is identified by its numeric disk address (for example, drive 2, cylinder 75, track 3, sector 13, etc.).

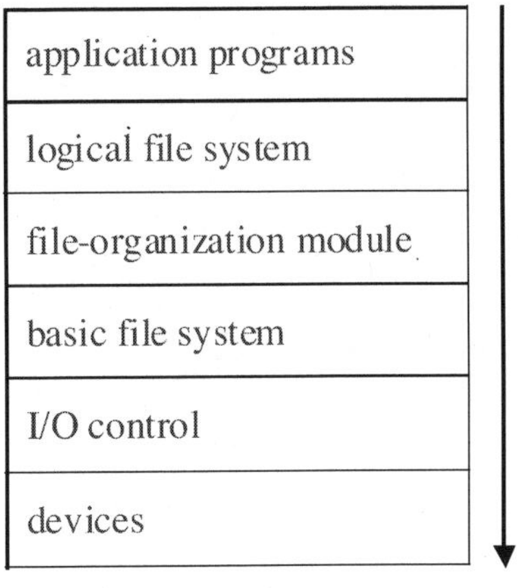

| application programs |
| logical file system |
| file-organization module |
| basic file system |
| I/O control |
| devices |

FIGURE 10.7. Layered file system

The *file-organization* module keeps information about the files and their logical and physical blocks. If the type of file allocation used and the location of the file is known, then the file-organization module can translate logical block addresses to physical block addresses for the basic file system to transfer.

The *logical blocks* of each file are numbered from 0 (or 1) through N. Because the physical blocks that contain the data usually do not match the logical numbers, a translation is required to locate each block. The free-space manager is also included in the file-organization module, whose task is to keep track of unallocated blocks and provide these blocks to the file-organization module when requested.

The metadata information is managed by the logical file system. All of the file-system structures (except the contents of the files) are included in the metadata. A *file-control block* (FCB) contains information about the file, which includes ownership, permissions, and the location of the file contents. The logical file system is also responsible for protection and security.

For a file-system implementation, when a layered structure is used, then the duplication of code is minimized. Sometimes both the I/O control and the basic file-system code can be used by multiple file systems. In such cases each file system can have their own logical file system and file system modules.

Most of the operating systems support more than one file system. For example, most CD-ROMs are written in the ISO 9660 format, a standard format agreed upon by CD-ROM manufacturers. In addition to removable-media file systems, each operating system has one disk-based file system (or more). UNIX uses the *UNIX file system* (UFS), which is based on the Berkeley Fast File System (FFS). Windows NT, 2000, and XP support disk file-system formats of FAT, FAT32, and KTFS (or Windows NT File System), as well as the CD-ROM, DVD, and floppy-disk file-system formats. Although Linux supports over forty different file systems, the standard Linux file system is known as the *extended file system*, with the most common version being ext2 and ext3. There are also distributed file systems, in which a file system on a server is mounted by one or more clients.

ALLOCATION METHODS

With the direct access nature of disks, the implementation of files becomes more flexible (in most cases many files are stored on the same disk). The major problem in file management is how the space for files will be allocated to get effective utilization of disk space and quick access of files. The three widely used disk space allocating methods are: contiguous, linked, and indexed. Each of these methods has its own advantages and disadvantages. Some systems (for example, Data General's RDOS) use all three allocation methods. Generally, a system will use only one particular method for all files.

Contiguous Allocation

In the contiguous allocation of files, each file occupies a set of contiguous blocks (addresses) on the disk. The addresses of a disk define a linear ordering on the disk. According to this linear ordering (assume that only one job is accessing the disk), accessing block n + 1 after block n does not require any head movement. But whenever head movement is required (it is required when needing to access the first sector of one cylinder after accessing the last sector of the previous cylinder), it will only move from one track to the next. Therefore, the required number of disk seeks for accessing contiguously allocated files is minimal, as is seek time when a seek is finally needed. Contiguous allocation of a file is defined by the disk address and length of the first block. If the file is h blocks long and starts at location n, then it occupies blocks n, n + 1, n + 2, . . ., n + h -1. The directory entry for each file indicates the address of the starting block and the length of the area allocated for this file. Contiguous allocation of files is shown in Figure 10.8.

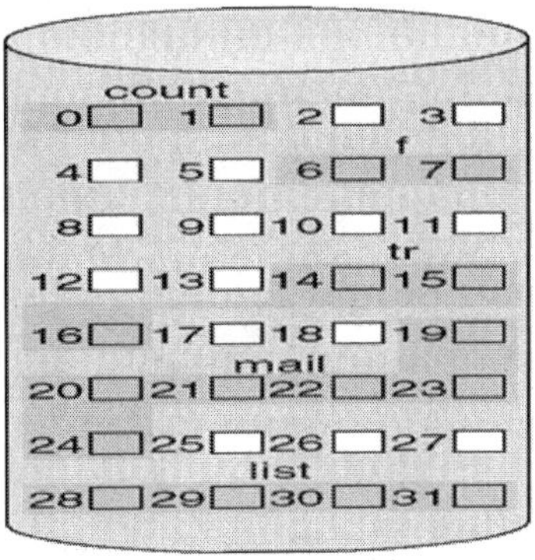

directory

file	start	length
count	0	2
tr	14	3
mail	19	6
list	28	4
f	6	2

FIGURE 10.8. Contiguous Allocation of Disk Space

To find the space for a new file is difficult in the contiguous allocation of files. If the file to be created is h blocks long, then the OS must search for h free contiguous blocks. First-fit, best-fit, and worst-fit are the most common strategies (as we have already studied) that are used to select a free hole from the set of available holes. According to the simulation results, both the first-fit and the best-fit strategies are better than worst-fit in terms of time and storage utilization. But in terms of storage utilization neither the first-fit nor the best-fit is best, but the first-fit strategy is generally faster.

All these algorithms also suffer from external fragmentation. As the files are allocated and deleted from the disk, then the free disk space is broken into small pieces. External fragmentation exists when enough disk space is available to satisfy a request, but this space is not available in a contiguous order; hence the storage is fragmented into a large number of small holes.

To determine how much disk space is required for a file is another problem that is associated with contiguous allocation. Whenever a new file is created, then the total amount of space it will require must be known and allocated. But how does the creator (program or person) come to know about the size of the file that has to be created? In some cases, it is easy to determine the size of a file (for example, copying an existing file), but generally the size of an output file may be difficult to estimate.

Linked Allocation

All the problems of contiguous allocation are solved by linked allocation. In linked allocation, each file is a linked list of disk blocks, where all the disk blocks may be scattered anywhere on the disk. In the linked allocation method, the directory contains a pointer to the first and the last block of the file. For example, as shown in Figure 10.9, a file that has five blocks starts with block 9 and then continues to block 16, then block 1, then block 10, and finally to block 25. Here each block contains a pointer to the next block.

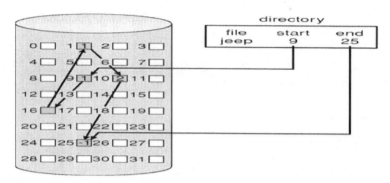

FIGURE 10.9. Linked allocation of disk space

To create a new file, a new entry is to be created in the directory. In the linked allocation method, each directory entry contains a pointer to the first disk block of the file. To denote an empty file, this pointer is initialized to nil (the end-of-list pointer value). The size field is also set to 0. A write operation takes place by finding the first free block, then this new block is written, and at last this new block is linked to the end of the file. To read a file, the blocks are read by following the pointers from block to block.

The linked allocation does not suffer from external fragmentation, and any of the free blocks can be used to satisfy a request. Therefore, the size of a file should not be declared at the time of file creation. The size of the file can grow as long as there are free blocks available.

Although linked allocation solves the external fragmentation and size declaration problems of contiguous allocation, it also has some disadvantages:

1. The major problem of linked allocation is that it supports the sequential access of files and is inefficient for direct access. For example, to find the jth block of a file, it will start at the beginning of that file and follow the pointers until the jth block is reached (each access to a pointer requires a disk read).
2. Another disadvantage of linked allocation is the space occupied by the pointers. For example, if a pointer requires 4 bytes out of a 512-byte block, then around 0.78% of the disk is being used for pointers, instead of for information.
3. The most severe problem of linked allocation is reliability. A bug in the operating system or a disk hardware failure may result in picking up the wrong pointer, and this error could in turn result in linking to another file.

A variation on the linked allocation is the use of a file-allocation table (FAT), shown in Figure 10.10. MS-DOS and OS/2 operating systems use this method of disk-space allocation. In FAT, a section of the disk at the beginning of each volume is set to one side to contain the table. This table contains one entry for each disk block and is indexed by the block number. The FAT can be used similarly as a linked list. In this the directory entry contains the block number of the first block of the file and the table entry which is indexed by block number contains the block number of the next block in the file. This chain continues until the last block of the table (i.e., end-of-file) is reached. Here all the unused blocks are indicated by a 0 table value. To allocate a new block to a file, the first 0-valued table entry has to be found and then the previous end-of-file value is replaced with the address of the new block. Hence, the 0 is then replaced with the end-of-file value.

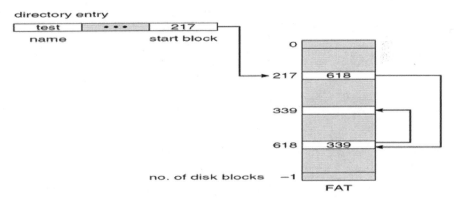

FIGURE 10.10. File Allocation Table

Indexed Allocation

We know that linked allocation is not efficient for direct access because the pointers to the blocks are scattered with the blocks all over the disk and they must be retrieved in order. Indexed allocation method is used to solve this problem by bringing all the pointers together into one location called the index block.

Each file contains its own index block, which is an array of disk-block addresses. The jth entry in the index block points to the jth block of the file. As shown in Figure 10.11, the directory contains the address of the index block. To find and read the jth block, the pointer is used in the jth index-block entry.

At the time of file creation, all pointers in the index block are set to nil. When the jth block is first written, a block is obtained from the free-space manager, and its address is put in the jth index-block entry.

Indexed allocation supports direct access of files without suffering from external fragmentation, because any free block on the disk may satisfy a request for more space.

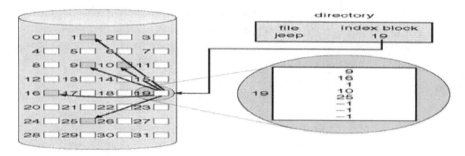

FIGURE 10.11. Indexed allocation of disk space

 Self-Quiz

Q11. Explain contiguous allocation of files.
Q12. Why is linked allocation better than contiguous allocation? Explain.
Q13. How is indexed allocation used for allocating files on disk?

FREE-SPACE MANAGEMENT

As we know that disk space is limited, it is necessary to reuse the space from deleted files for new files. A *free-space list* is maintained to keep track of free disk space. This list records all free disk blocks, that is, those blocks that are not allocated to some file or directory. When creating a file, the free-space list is searched for the required amount of space and allocates that space to the new file. This allocated space is then removed from the free-space list and when a file is deleted, its disk space is added to the free-space list.

Bit Vector

The free-space list is commonly implemented as a *bit map* or *bit vector*. In this method each block is represented by 1 bit. If the block is free, the bit is 1, and if the block is allocated, the bit is 0. For example, when considering a disk where blocks 1, 3, 4, 5, 7, 9, 10, 11, 12, 13, 16, 18, 25, 26, and 28 are free and the rest of the blocks are allocated, the free-space bit map would be
0101110101111100101000000011010000 . . .
This method is simple and efficient in finding the first free block or n consecutive free blocks on the disk. In this technique, the first non-0 word, that is, 1 bit, is scanned, which represents the location of the first free block. The block number can be calculated by:
(Number of bits per word) x (number of 0-value words)+offset of first 1 bit. The bit vectors are inefficient unless the entire vector is kept in main memory.
For small disks it is possible to keep the entire vector in the main memory, but the same is not necessarily possible for large disks.

Linked List

Free-space management can also be done through a linked list, where all the free disk blocks are linked together, by keeping a pointer to the first free block in a special location on the disk and caching it in memory. This first block contains a pointer to the next free disk block, and so on. For example (considering the previous section example), a pointer is kept to block 1 as the first free block. Block 2 would contain a pointer to block 3, which would point to block 4, which would point to block 5, which would point to block 7, and so on. This scheme is not efficient because to traverse the list, each block must be read, which requires considerable I/O time. Although traversing a free list is not a frequent action. Generally, the first free block is required by the operating system so that it can allocate that block to a file. The FAT (File Allocation Table) method includes free-block accounting in the allocation data structure. No other separate method is required.

Grouping

In this approach of free-space management, the addresses of *n* free blocks are stored in the first free block. Among these blocks, the first n-1 blocks are actually free. The last block contains the addresses of other n free blocks, and so on. As compared to the liked list approach, in this method the addresses of a large number of free blocks can be found quickly.

Counting

It is known that several contiguous blocks may be allocated or freed simultaneously, when the space is allocated with the contiguous-allocation algorithm or through clustering. Therefore, rather than keeping a list of *n* free blocks, one should keep the address of the first free block and the number *n* of free contiguous blocks that follow the first block. Now each entry in the free-space list consists of a disk address and a count. Even though each entry requires more space than a simple disk address, the overall list will be shorter, as long as the count is generally greater than 1.

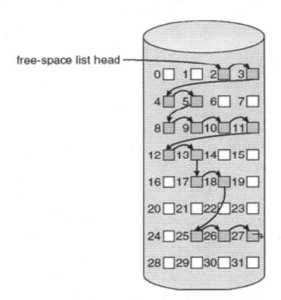

FIGURE 10.12. Linked free-space list on disk

PROTECTION

It is necessary to keep the information (that is stored in a computer system) safe from physical damage (called reliability) and improper access (called protection). Reliability can be generally provided by making duplicate copies of the files. Protection can be provided in many ways. For a single user system, protection can be provided by physically removing the floppy disks. However, in a multiuser system, other mechanisms are needed.

Types of Access

Because the files are being accessed by other users, protection is required. The systems that do not access the files of other users do not require protection. Therefore, the complete protection of a system can be provided by prohibiting access. An alternative method is to provide free access with no protection. Since both the approaches are not feasible, *controlled access* is required.

Controlled access can be provided to the protection mechanism by limiting the types of file access. Access can be granted or denied depending on some factors. The type of access requested is one of them. Different types of operations that may be controlled are:

- **Read:** Read from the file.
- **Write:** Write or rewrite the file.
- **Execute:** Load the file into memory and execute it.
- **Append:** Write new information at the end of the existing file.
- **Delete:** Delete the file and free its space. This space can be reused.
- **List:** List the name and attributes of the file.

Other operations, like renaming, copying, and editing the file, may also be controlled. For most of the systems, these higher-level functions may be implemented by a system program that makes lower-level system calls, and the protection is provided at the lower level only. For example, to copy a file may simply be implemented by a sequence of read requests. Here, the user with read access can also cause the file to be copied, printed, and so on.

Many of the protection mechanisms have been proposed, and they all are appropriate for their intended application and each of them has their own advantages and disadvantages.

Access Control

For protection the most common approach is to make access totally dependent on the identity of the user. Different types of files or directories can be accessed by different users. To implement the identity-dependent access, each file and directory is associated with an Access Control List (ACL). ACL specifies the user names and the types of the access allowed for each user. When a user requests access to a particular file, the operating system then checks the access control list associated with that file. If that user's identity is available in the list for the requested access, the access is allowed. Otherwise, a protection violation occurs.

The advantage of this approach is that it enables complex access methodologies. The length of the access list is a big problem. If everyone is allowed to read a file, then all the users must be listed with read access. This technique has two adverse effects:

■ To construct such a list is a tedious and difficult task, especially when the list of users in the system is not known in advance.
■ The fixed-size directory entry now needs to be of variable size. This results in more complicated space management.

These problems can be solved by using the condensed version of the access list. To condense the length of the ACL, most of the systems provide three classifications of users (in connection with each file):

■ **Owner:** Owner is the user who created the file.
■ **Group:** A group is a set of users who are sharing the file and need similar access.
■ **Universe:** All the other users in the system constitute the universe.

The most common and easy-to-implement approach is to combine access control lists with the more general owner, group, and universe access control scheme.

Other Protection Approaches

Another approach for protection is to associate a password with each file. As the password is required to access the computer system, similarly, access to each file can also be controlled by the password. If the random passwords are chosen and are changed regularly, then this scheme may be effective in limiting the access to a file. The use of passwords also has some disadvantages:

1. The number of passwords that a user desires to remember may become large, making this scheme impractical.
2. If only one password is used for all the files, then once it is exposed, all files are accessible.

To deal with this problem some of the systems (for example, TOPS-20) allow a user to associate a password with a subdirectory, rather than with an individual file. The IBM's VM/CMS (International Business Machines Virtual Machine/Cambridge Monitor System) operating system allows three passwords for a minidisk, one each for read, write, and multi-write access.

(⦿) Self-Quiz

Q14. What is file protection? Explain.
Q15. How can the file be protected?

CASE STUDIES

FAT32 File System

In this section we will discuss the FAT32 file system that is included with Microsoft Windows XP. An updated version of the FAT file system, that is, FAT 32, is included in Windows XP. The FAT32 file system allows a default cluster size of 4 KB, and includes support for EIDE (Enhanced Integrated Drive Electronics) hard disk sizes larger than 2 gigabytes (GB).

FAT32 Features

FAT32 has the following features:

- FAT32 supports drives of up to 2 terabytes in size.
- Space utilization of FAT32 is more efficient. FAT32 uses smaller clusters (i.e., 4 KB clusters for drives of up to 8 GB in size), so 10–15% more disk space can be used efficiently.
- FAT32 is robust. In this the root folder can be relocated and the backup copy of the file allocation table is used instead of the default copy. The boot record on FAT32 drives is expanded to include a backup copy of critical data structures. For that reason, FAT32 drives are less vulnerable to a single point of failure.
- FAT32 is flexible. The root folder on a FAT32 drive is an ordinary cluster chain; therefore, it can be located anywhere on the drive.

FAT32 Compatibility Considerations

To maintain the compatibility with existing programs, networks, and device drivers, FAT32 was implemented with a slight change to the existing Windows architecture, internal data structures, Application Programming Interfaces (APIs), on-disk format, and so on. To store the cluster values, only 4 bytes are now used. Hence, many internal and on-disk data structures and published APIs have been revised or expanded. In some of the cases, the existing APIs will not work on FAT32 drives. Most programs will remain unaffected by these changes.

MS-DOS blocks device drivers (for example, Aspidisk.sys), and disk tools will need to be revised to support FAT32 drives.

All of Microsoft's disk tools have been revised to work with FAT32. Also, Microsoft is working with leading device driver and disk tool manufacturers to support them in revising their products to support FAT32.

Note: A FAT32 volume cannot be compressed by using Microsoft DriveSpace or DriveSpace 3.

Dual-Boot Computers

FAT 32 volumes can be accessed by Windows XP, Windows 2000, Windows Me, Windows 98, and Windows 95 OSR2. The original version of Windows 95, i.e., MS DOS, and Windows NT 4.0 do not recognize FAT32 partitions, and hence are unable to boot from a FAT32 volume.

Note: FAT32 volumes cannot be accessed properly if the computer is started by using another operating system (for example, a Windows 95 or MS-DOS boot disk).

Creating FAT32 Volumes

In Windows XP the Disk Management snap-in is a tool for managing hard disks and the volumes or partitions that they contain. Disk management can be used to create new FAT32 volumes or format an existing volume to use the FAT32 filesystem. FAT32 can also be used by formatting the basic and dynamic volumes.

Creating a FAT32 Partition or Logical Drive: To create a new FAT32 partition or logical drive in Windows XP, the following steps are to be taken:

- Log on as an Administrator or as a member of the Administrators group.
- Now click *Start*, then right-click *My Computer*, and then click *Manage*.
- In the console tree, click *Disk Management*.
- In the Disk Management window, perform one of the following:

  - For creating a new partition, right-click unallocated space on the basic disk where you want to create the partition, and then click *New Partition*. OR
  - For creating a new logical drive in an extended partition, right-click free space on an extended partition where you want to create the logical drive, and then click New Logical Drive.

- Now click *Next* in the New Partition Wizard.
- Click on the type of partition that you want to create (i.e., *Primary partition*, *Extended partition*, or *Logical drive*), and then click *Next*.
- The size of the partition is to be specified in the **Partition size in MB** box, and then click *Next*.
- Now assign a drive letter or drive path to the new partition or logical drive, and then click *Next*.
- Click **Format this partition with the following settings**, and then perform the following actions:

  - In the **File system** box, click *FAT32*.
  - In the **Volume label** box, type a name for the volume.
    The disk allocation unit size can also be changed if required, or a quick format can be performed if specified.

- Click *Next*.
- Make it confirm that the selected options are correct, and then click *Finish*.

In the appropriate basic disk in the Disk Management window, the new partition or logical drive is created and appears.

Formatting an Existing Volume to Use FAT32: The following steps are taken to format a volume:

- Click *Start*, then right-click *My Computer*, and then click *Manage*.
- Click Disk Management in the *console tree*.
- Right-click the volume in the Disk Management window that is to be formatted (or reformatted), and then click *Format*.
- Do the following in the *Format dialog box*:

    - In the **File system** *box*, click FAT32.
    - In the **Volume label** *box*, type a name for the volume.
      The disk allocation unit size can also be changed if required, or a quick format can be performed if specified.

- Click *OK*.
- When you are prompted to format the volume, click *OK*.
- Now the format process starts.

New Technology File System (NTFS)

NTFS was designed by Microsoft as a successor to FAT32. It was introduced with the Windows NT operating system, and therefore it is named NTFS. This file system supports 64-bit volume sizes. Disk space is allocated in *extents* on top of *clusters*. The cluster size is a multiple of the disk's block size, usually $n \times 512$ bytes. For today's disk the common cluster size in NTFS is 4KB. NTFS provides *journaling* for reliability and *file-based data compression* to reduce the space that a file uses on the disk.

The Master File Table (MFT) used in NTFS is similar to an inode. For all files in the system, the MFT contains file records (Microsoft uses this term for an inode). These records are organized in a B-Tree structure in place of a linear array, which is used by most Unix-derived file systems (UFS, FFS, ext2, ext3, etc.). In Unix-derived file systems, the inode area is a bunch of pre-allocated clusters that are separate from file data, whereas the MFT (used in Windows NT) itself is a file and can grow like any other file. It is a special file known as a *metafile*. It can be allocated and managed like any other file.

Other metafiles (special files) that can be handled just like other files are:

- Log file
- Volume file
- Attribute definition file
- Bitmap file (free blocks)
- Boot file (located in low-numbered consecutive blocks so the boot loader can read them)
- Bad cluster file
- Root directory

The volume bitmap is a file which identifies free space within the volume. This file can grow dynamically; hence, it allows the size of a volume to be expanded dynamically.

Each entry of a file record in MFT contains a set of typed attributes, which include the file name, creation date, and permissions. Some attributes may have multiple instances. For example, a file name may also have an alternate MS-DOS-compatible 8.3-character name.

Usually a file requires only one record. If more records are needed, then the first record for the file will store the location of other file records. Hence an "attribute list" attribute is added, which identifies the location of all other attribute records for the file.

Unlike the other file system, small files and directories (1,500 bytes or smaller) are completely contained within the file's MFT record. As a result, no additional data blocks are required to allocate these small files and directories. Also, no additional data blocks are required to access the contents of the files.

If a file is very large, then no data will be stored within the file record. In such cases, the file record data attribute contains pointers to the file data. This collection of pointers is a list of *extents*. Here each pointer contains:

1. Starting cluster number in the file (cluster offset in the file).
2. Cluster number on the disk where this extent is located.
3. Cluster count: number of clusters in this extent.

As a file grows, additional extents are allocated if the last extent cannot be expanded. If the file record (MFT entry) runs out of space, another MFT entry is allocated.

Under NTFS a file may have multiple data streams associated with it. A stream gets selected when the file is opened with a specific stream name (e.g., "filename:streamname"). Generally, this is not specified and the default, that is, the main stream, is used. Additional streams are usually lost when the file is copied to other file system types, such as FAT32 on a USB flash drive.

An NTFS directory is a file that contains file names and a reference to the file. The reference identifies the file's location in the MFT table. If the directory is small, it is stored in the MFT entry. But if it is large, then it is stored in other clusters, because the files are too large to fit into an MFT entry.

To make searching possible through large directories, the directory file is organized as a sorted B+ tree. In this case the MFT entry contains the top node of the B+ tree and identifies the extents and offsets where the directory entries live. Directories also contain data such as file size, last access time, and last modification time. These data are redundant since they are present in the MFT entry. This is done to optimize some directory listings. Presenting the user with these common attributes does not require reading the file record for each file in the directory.

NTFS allows the files to be compressed. Compression of either individual files, files within a directory, or all files within the volume may be selected. Compression is handled within the file system. It allows higher levels of the operating system to be unaware that it is happening. When a file is being read, NTFS reads that portion into memory and decompresses it. Files are compressed in 64 KB chunks. If the compression results in a size of 60 KB or less, then the file system has saved at least one 4 KB cluster.

EXT2/EXT3/EXT4 File Systems of Linux

Ext2, Ext3, and Ext4 are the file systems that are created for Linux. In this section we will discuss the following:

■ High level differences between these file systems.
■ How to create these file systems.
■ How to convert one file system type to another type.

Ext2

Ext2 or Second Extended File System was introduced in 1993 and was developed by Remy Card. Ext2 does not include a journaling feature. Ext2 is recommended on flash drives and USB drives, as it does not need to do the overhead of journaling. The maximum individual file size can be from 16 GB to 2 TB. The overall ext2 file system size can be from 2 TB to 32 TB.

Ext3

Ext3 or Third Extended File System was introduced in 2001 and was developed by Stephen Tweedie. Starting from Linux Kernel 2.4.15, ext3 was available. The advantage of ext3 is that it allows journaling. In the file system, journaling has a dedicated area where all the changes are tracked. Journaling lowers the possibility of file system corruption when the system crashes. The maximum individual file size can be from 16 GB to 2 TB, and the overall ext3 file system size can be from 2 TB to 32 TB. The three types of journaling available in the ext3 file system are:

■ **Journal:** Content and metadata are saved in the journal.
■ **Ordered:** Only the metadata is saved in the journal. By default, metadata are journaled only after writing the content to disk.
■ **Writeback:** Only the metadata is saved in the journal. Metadata might be journaled either before or after writing the content to the disk.

The ext2 file system can be directly converted to the ext3 file system (without backup/restore).

Ext4

Ext4 or Fourth Extended File System was introduced in 2008. Ext4 was available starting from Linux Kernel 2.6.19. It supports a huge individual file size and overall file system size. The maximum individual file size can be from 16 GB to 16 TB and the overall maximum Ext4 file system size is 1 EB (Exabyte). Where 1 EB = 1024 PB (petabyte) and 1 PB = 1024 TB (terabyte).

The Ext4 directory can contain a maximum of 64,000 subdirectories, whereas the Ext3 directory can contain maximum of 32,000. An existing Ext3 file system can be mounted as an Ext4 file system, without upgrading it.

In Ext 4 several other new features are introduced, such as: multiblock allocation, delayed allocation, journal checksum, fast fsck, and so on. In Ext4, there is also an option for turning "OFF" the journaling feature. Once you've partitioned your hard disk using the *fdisk* command, use *mke2fs* to create either an Ext2, Ext3, or Ext4 file system.

Create an ext2 file system:
    mke2fs /dev/sda1

Create an ext3 file system:
    mkfs.ext3 /dev/sda1
    (or)
    mke2fs –j /dev/sda1

Create an ext4 file system:
    mkfs.ext4 /dev/sda1
    (or)
    mke2fs -t ext4 /dev/sda1

Converting ext2 to ext3
    For example, if you are upgrading /dev/sda2 that is mounted as /home, from ext2 to ext3, do the following:
    umount /dev/sda2
    tune2fs -j /dev/sda2
    mount /dev/sda2 /home

Converting ext3 to ext4
    If you are upgrading /dev/sda2 that is mounted as /home, from ext3 to ext4, do the following.
    umount /dev/sda2
    tune2fs -O extents,uninit_bg,dir_index /dev/sda2
    e2fsck -pf /dev/sda2
    mount /dev/sda2 /home

SUMMARY

A file is an abstract data type. It is a sequence of logical records. A logical record may be a byte, a line, or a more complex data item.

In a file system each device keeps a device directory that lists the location of the files on the device. In a multiuser system a single-level directory causes a naming problem, since each file has a unique name. This problem is solved by two-level directory, where for each user a separate directory is created. The directory lists the files by name and includes the location of the file on the disk, length, type, owner, time of creation, time of last use, and so on. The generalization of a two-level directory is a tree-structured directory that allows a user to create subdirectories to organize files. Acyclic-graph directory structures enable the users to share subdirectories, and file searching and deletion becomes complicated with this. Sharing of the files and directories is completely flexible with a general graph structure, but it sometimes requires garbage collection to recover the unused disk space.

The space on the disk can be allocated to the main memory in three ways, contiguous, linked, or indexed allocation. Contiguous allocation suffers from external fragmentation. Linked allocation is inefficient for direct access. Indexed allocation may require significant overhead for its index block. All these algorithms can be optimized in various ways. To increase flexibility and to decrease external fragmentation, contiguous space can be enlarged. To increase the throughput and to reduce the number of index entries, indexed allocation can be done in clusters of multiple blocks.

Since files are the main information, they need to be protected. For each type of access, such as read, write, execute, append, delete, list, and so on, access to files can be controlled individually. Files can be protected by using passwords, by access lists, or by other techniques. Finally, case studies on file systems of FAT32, NTFS, and EXT2/EXT3/EXT4 are discussed in brief.

POINTS TO REMEMBER

- A disk can be considered as a linear sequence of fixed-size blocks.
- A file is a named collection of related information.
- A directory contains the name of the file and its unique identifier.
- The two levels of internal tables that are used by the operating system are a per-process table and a system-wide table.
- File locks allow one process to lock the file and prevent its access from other processes.
- To allocate the complete file into blocks rather than bytes is called internal fragmentation.
- In the sequential access method, processing of the information that is stored in a file must be done in order.
- In a direct access method, disks allow random access to any block of a file.
- The part of the disk that holds the file system is referred to as the volume.
- In the single-level directory structure, all the files are contained in the same directory.
- In a two-level directory, each user has a separate user file directory.
- The path name can be of two types, absolute and relative.
- Linked allocation supports sequential access of files and is inefficient for direct access.
- In indexed allocation, each file contains its own index block.

REVIEW QUESTIONS

Q1. Explain the file system and its function and how system calls are related to a file system.

Q2. Write a short note on the direct access method of files.

Q3. What are the different file organizations? Discuss access mechanisms.

Q4. Explain the following:
   a. Directory system
   b. File protection

Q5. Discuss various types of file directory structure with a relevant diagram.

Q6. Describe the various file allocation methods with their relative advantages and disadvantages.

Q7. What are the various file access methods? Explain the sequential access method. How can it be simulated on a direct access file?

Q8. Describe file system implementation and directory system implementation in detail.

Q9. What is a directory? Explain any two ways to implement the directory.

Q10. Write short notes on:
   a. Sequential files
   b. Indexed files

CHAPTER 11

I/O SYSTEMS

Purpose of This Chapter

This chapter will explore the structure of the I/O subsystem of an operating system, discuss the principles of I/O hardware and its complexity, and provide the details of the performance aspects of I/O hardware and software.

INTRODUCTION

Any computer can perform mainly two jobs, that is, Input/Output and processing. In most cases, the main job of a computer is I/O, and the processing is only incidental. For example, when a user browses a Web page or tries to edit a file, his or her main concern is to read or enter some information, not to compute an answer.

In a computer, the operating system's role is to manage and control I/O operations and I/O devices. In this chapter, we first describe the basics of I/O hardware. After that we will discuss the I/O services provided by the operating system and the examples of these services in the application I/O interface. At last, we explain how the operating system bridges the gap between the hardware interface and the application interface.

OVERVIEW

The major concern of an operating system is to manage the devices that are connected to the computer. As I/O devices vary so broadly in their function and speed, various methods are required to control them. These methods form the I/O subsystem of the kernel.

I/O subsystems exhibit two conflicting trends: (a) Increasing standardization of software and hardware interfaces is seen, which helps to incorporate newly improved devices into existing computers and operating systems; (b) The development of an entirely new type of I/O devices is increasing, and it is a challenge to incorporate them into computers and operating systems (because some of the new devices are very different from the previous devices).

Basic I/O hardware elements (e.g., ports, buses, device controllers, etc.) accommodate a wide range of I/O devices. To handle a particular device or category of similar devices, the kernel of an operating system is structured to use device driver modules.

I/O HARDWARE

Computers can operate on various types of devices. Categories of I/O devices could be storage, transmission, user interface, and so on. Although there are a variety of I/O devices, only a few concepts are required to understand how these devices are attached and how the hardware can be controlled by the software.

Devices can communicate with a computer system by sending signals either through the wires or through the air. Ports are used to connect devices with the computer, for example, a serial port. If the device uses a common set of wires, then the connection is called a bus. A bus is a set of wires and a strictly defined protocol that specifies a set of messages that can be sent over the wires. Figure 11.1 shows the three of the four types of bus that are commonly found in a modern PC (Personal Computer).

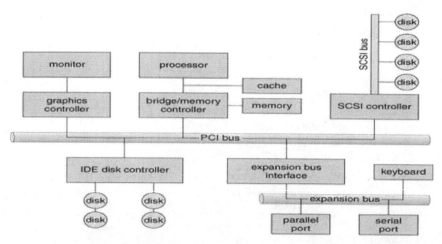

FIGURE 11.1. A Typical PC Bus Structure

1. The *PCI (Peripheral Component Interconnect) bus* connects the fast devices to the processor memory subsystem.
2. The *expansion bus* connects slower devices, such as the keyboard and serial and parallel ports.
3. The *SCSI (Small Computer System Interface) bus* connects a number of SCSI devices to a common SCSI controller. A controller is basically a collection of

electronics that can operate a device, a bus, or a port. A serial port controller is a simple device controller. In the computer, it is a single chip that controls the signals on the wires of a serial port. A SCSI bus controller is complex, because the SCSI protocol is complex. The SCSI bus controller is implemented as a separate circuit board that is plugged into the computer. It contains a processor, microcode, and some private memory that enable it to process the SCSI protocol messages.

4. If there are three devices X, Y, and Z, then in a *daisy chain bus* arrangement device X has a cable that plugs into device Y, and device Y has a cable that plugs into device Z, and device Z plugs into a port on the computer. In a daisy chain arrangement only one device is directly connected to the host.

The controller contains one or more than one register for data and control signals. To communicate with the controller, the processor reads and writes bit patterns on the registers. One way in which this communication can take place is through the use of registers associated with each port. The data registers can be of 1 to 4 bytes in size. Typically an I/O port may include the following four registers:

1. The *data-in register* is read by the host to receive input from the device.
2. The *data-out register* is written by the host to send output.
3. The *status register* has bits that can be read by the host to determine the status of the device, that is, whether the status is idle, ready for input, busy, error, transaction complete, and so on.
4. The *control register* has bits that are written by the host to issue a command or to change the settings of the device, such as parity checking, word length, full- versus half-duplex operation, and so on.

Table 11.1 shows some of the most common ranges of the I/O port address.

TABLE 11.1. Device I/O Port Locations on PCs (Partial)

I/O address range (hexadecimal)	Device
000-00F	DMA controller
020-021	interrupt controller
040-043	Timer
200-20F	game controller
2F8-2FF	serial port (secondary)
320-32F	hard-disk controller
37B-37F	parallel port
3D0-3DF	graphics controller
3F0-3F7	diskette-drive controller
3F8-3FF	serial port (primary)

An alternative method through which communication takes place with devices is *memory-mapped I/O*. In this technique a certain portion of the address space of any processor is mapped to the device, and communication takes place by reading and writing directly to or from those memory areas. The memory-mapped I/O technique is appropriate for those devices that can move large quantities of data quickly, such as graphics cards, and those that still use registers to control the information, such as setting the video mode. A problem arises in the memory-mapped I/O technique when a process is allowed to write directly to the address space used by a memory-mapped I/O device.

Some systems use both techniques. For instance, PCs use I/O instructions to control some devices and memory-mapped I/O to control others.

Polling

The idea of *handshaking* for interaction between the host and a controller could be simple. Device handshaking involves polling. For example, the host writes output through a port, coordinating with the controller by handshaking as follows:

1. The host constantly checks the *busy bit* on the device until it becomes clear.
2. The host writes a byte of data into a data-out register, and sets the *write bit* in the command register.
3. The host sets the *command ready bit* in the command register to notify the device of the pending command.
4. When the device controller sees the command-ready bit set, it first sets the busy bit.
5. Then the device controller reads the command register, sees the write bit set, reads the byte of data from the data-out register, and outputs the byte of data.
6. The device controller then clears the *error bit* in the status register and the command-ready bit, and finally clears the busy bit, signaling the completion of the operation.

Polling can be very fast and efficient if both the device and the controller are fast and if there is significant data to be transferred. It becomes inefficient, however, if the host must wait a long time in the busy loop waiting for the device, or if frequent checks need to be made for data that is infrequently there.

Interrupts

Interrupts are used because they allow the devices to report to the CPU that when they have data to be transferred or when an operation is to be completed, it also allows the CPU to perform other duties when the CPU is not required immediately for a data transfer. The working mechanism for basic interrupt is as follows:

■ The hardware of the CPU has a wire which is called an interrupt-request line, and this line is sensed after the execution of each instruction. Whenever the CPU senses that the controller has declared a signal on the interrupt request line, then:

• By declaring a signal on the interrupt request line an interrupt is raised through a device's controller.
• A state save is performed through the CPU, and the CPU also transfers control to the *interrupt handler* routine in memory at a fixed

address; that is, interrupts are caught and the interrupt handler is dispatched by the CPU.

■ The cause of the interrupts is determined by the interrupt handler. The interrupt handler also performs necessary processing; it performs a state restore, and returns control to the CPU, executing a return from interrupt instruction.

The interrupt-driven I/O procedure is shown in Figure 11.2. The interrupt-driven I/O cycle is performed and controlled by some of the prime steps which are discussed below.

■ **Step 1:** At Step 1, the device driver initiates I/O.

■ **Step 2:** After initiating I/O, the cycle reaches Step 3.

■ **Step 3:** The tasks, such as input ready, output complete, or error generates interrupt signal all are checked at this step.

■ **Step 4:** Then the cycle goes to Step 4. At this step, the CPU receiving the interrupt transfers control to the interrupt handler. An interrupt handler (Interrupt Service Routine or ISR) is a callback function in an operating system whose execution is triggered by the response of an interrupt.

■ **Step 5:** At this stage, the interrupt handler is used to process data as the reception of an interrupt.

■ **Step 6:** The cycle is completed. Here, the CPU resumes processing of the interrupted task.

■ The procedure shown in Figure 11.2 is suitable for a simple interrupt-driven I/O. But in modern computing there are the following three requirements that complicate the situation:

  • The requirement to defer interrupt handling during critical processing.

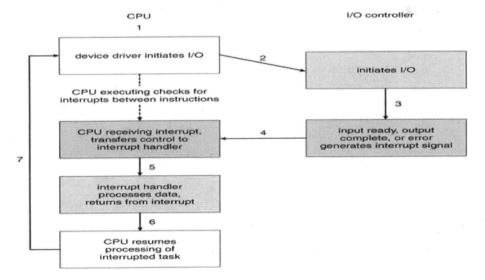

FIGURE 11.2. Interrupt-driven I/O cycle

- The requirement of an efficient way to determine which interrupt handler to be invoked, without polling all devices to see which of the devices need attention.
- To differentiate between high-priority and low-priority interrupts, there is a requirement of multi-level interrupts for a proper response.

■ In modern computer hardware these three issues are handled with the help of *interrupt controller* hardware.

- Most CPUs have two interrupt request lines, that is, non-maskable and maskable interrupt. *Non-maskable* interrupts are reserved for critical error conditions, such as unrecoverable memory errors, whereas *maskable* interrupts can be temporarily ignored by the CPU during critical processing, and these interrupts are used by the device controllers to request service.
- An address is accepted by the interrupt mechanism, where the address is usually a number which selects a specific interrupt handling routine from a small set of numbers. This address is offset into a table known as the *interrupt vector*. The table holds the addresses of the routines prepared to process specific interrupts.
- As we know that computers are having more devices, the number of possible interrupt handlers is more than the address elements in the interrupt vector. This problem can be solved by the technique of interrupt chaining, where each of the elements of the interrupt vector point to the head of the list of interrupt handlers. The handlers are called one by one from the corresponding list whenever an interrupt is invoked. This process continues until the handler is not found to provide its services. This structure is cooperation between the overhead of a huge interrupt table and the inefficiency of dispatching to a single interrupt handler.
- Table 11.2 represents the interrupt vector for Intel Pentium. Interrupts 0 to 31 are non-maskable and are reserved for serious hardware and other errors. Maskable interrupts that include normal device I/O interrupts begin at interrupt number 32.
- In modern computers interrupt hardware also supports *interrupt priority levels*, which allows the systems to mask off only lower-priority interrupts while a high-priority interrupt is being serviced, or allows a high-priority signal to interrupt the processing of a low-priority interrupt.

■ The system determines at boot time which devices are available, and installs the corresponding handler addresses into the interrupt vector.

■ During I/O, device controllers signal the completion of commands or availability of input data or detection of failure via interrupts.

■ A wide variety of exceptions, such as dividing by zero, access of invalid memory, or attempting to access kernel mode instructions can be handled via interrupts.

TABLE 11.2. Intel Pentium Processor Event-Vector Table

vector number	description
0	divide error
1	debug exception
2	null interrupt
3	breakpoint
4	INTO-detected overflow
5	bound range exception
6	invalid opcode
7	device not available
8	double fault
9	coprocessor segment overrun (reserved)
10	invalid task state segment
11	segment not present
12	stack fault
13	general protection
14	page fault
15	(Intel reserved, do not use)
16	floating-point error
17	alignment check
18	machine check
19–31	(Intel reserved, do not use)
32–255	maskable interrupts

- The mechanism of interrupts can also implement time slicing and context witches.
- System calls are implemented by software interrupts, which are also known as *traps*.
- The interrupt mechanism can also be used to schedule the activities for optimal performance and to control the kernel operations.

Direct Memory Access

For a device that transfers a large quantity of data, for example, disk controllers, it is a waste to use a general-purpose processor to lookout status bits and to transfer data in and out of controller registers 1 byte at a time; this process is known as *Programmed I/O (PIO)*.

Instead of overloading the main CPU with PIO, some of this work can be offloaded to a special purpose processor, which is known as the *Direct Memory Access (DMA) Controller*. The DMA transfer is initiated when the DMA command block is written into memory by the host.

The DMA command block comprises a pointer to the location where the data is stored, that is, source, a pointer to the location where the data is to be transferred, that is, destination, and the number of bytes to be transferred. The DMA controller first handles the data transfer, and after the completion of the transfer it interrupts the CPU.

In modern computers a simple DMA controller is a standard component, and many of the bus-mastering I/O cards contain their own DMA hardware.

The procedure of handshaking between the DMA controllers and their devices takes place through two wires, that is, the DMA-request and DMA-acknowledge wires. When the DMA transfer is going on, the CPU cannot access the PCI bus, but it can access its internal registers and primary and secondary caches. DMA can be done in terms of either the physical addresses or the virtual addresses (mapped to physical addresses). The approach where DMA is done through virtual addresses is known as *Direct Virtual Memory Access (DVMA)*. This approach allows direct data transfer from one memory-mapped device to another without using the main memory chips.

To speed up operations, DMA can be directly accessed by user processes, but for security and protection reasons this is prohibited by modern computers. Figure 11.3 shows the entire DMA process. The following are the required steps to perform the DMA process:

- **Step 1:** The device driver is told to transfer and relocate disk data to the buffer at address X. A device driver is a computer program that controls a particular type of device that is attached to a computer. It provides a software interface to enable operating systems to access hardware functions without needing to know precise details of the hardware being used.
- **Step 2:** The device driver tells the disk controller to transfer C bytes from the disk to the buffer at address X. The disk controller is the circuit which enables the CPU to communicate with a hard disk, floppy disk, or other kind of disk drive.
- **Step 3:** The disk controller initiates the DMA transfer. Most hardware systems use DMA, including disk drive controllers, graphics cards, network cards, and sound cards.
- **Step 4:** The disk controller sends each byte to the DMA controller. With DMA, the CPU initiates the transfer, performs other operations while the transfer is in progress, and receives an interrupt from the DMA controller when the operation is done. This feature is useful any time the CPU cannot keep up with the rate of data transfer or where the CPU needs to perform useful work while waiting for a relatively slow I/O data transfer.
- **Step 5:** The DMA controller transfers bytes to buffer X, increasing memory address C and decreasing C until C = 0.
- **Step 6:** When C = 0, DMA interrupts the CPU to signal transfer completion.

A CPU cache is used by the CPU to reduce the average time to access data from the main memory as shown in Figure 11.3. The Peripheral Component Interconnect or PCI bus as shown in Figure 11.3 provides a bus architecture that supports peripherals and devices like hard disk drives, networks, and so on.

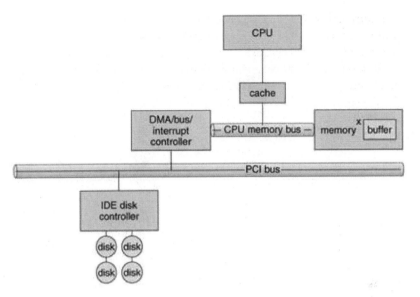

FIGURE 11.3. DMA process

One of the prime features of the PCI bus is that all data transfers in the PCI bus take place according to a system clock

Self-Quiz

Q1. What is polling?
Q2. Explain interrupt.
Q3. What do you understand about DMA?

APPLICATION OF I/O INTERFACE

This section discusses the structuring techniques and the interfaces for the operating system that enable the I/O devices to be treated in a standard, uniform way.

■ A variety of different devices can be accessed by user application through software layering, and through the encapsulation process of all of the device-specific code into device drivers, while the application layers are presented with a common interface for all the devices. Figure 11.4 represents how the I/O-related portions of the kernel are structured in software layers.

■ Devices can differ in many different dimensions. This differentiation is shown in Table 11.3.

Most of the devices are characterized as block I/O, character I/O, memory mapped file I/O, or network devices. Some special devices are characterized as clock and timer.

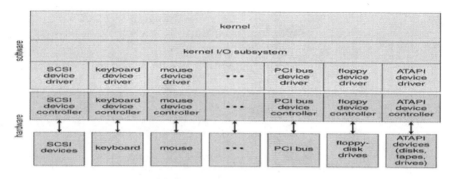

FIGURE 11.4. A kernel I/O structure

TABLE 11.3. Characteristics of I/O Devices

aspect	variation	example
data-transfer mode	character block	terminal disk
access method	sequential random	modem CD-ROM
transfer schedule	synchronous asynchronous	tape keyboard
sharing	dedicated sharable	tape keyboard
device speed	latency seek time transfer rate delay between operations	
I/O direction	read only write only road—write	CD-ROM graphics controller disk

Block and Character Devices

One block at a time is accessed by *block devices*. The operations supported by block devices are read(), write(), and seek().

■ To access the blocks directly on a hard drive is called *raw I/O*. Using this, certain operations can be sped up by bypassing the buffering and locking that are conducted by the OS.

■ An alternative method to access the blocks is *direct I/O*. Direct I/O uses the normal file system access, but the buffering and the locking operations are disabled.

■ On the top of the block-device drivers, the memory-mapped file I/O can be layered.

■ Instead of reading the entire file, it is mapped to a specific range of memory addresses, and then paged into memory according to the requirement by using the virtual memory system.

■ Access to the file is accomplished through normal memory accesses, instead of using the read() and write() system calls. This approach can be used for executable program code.

One byte at a time can be accessed by *character devices*. The operations that are supported by character devices are get() and put().

Network Devices

Network access is fundamentally different from local disk access; therefore, most of the systems provide a distinct interface for network devices.

Socket interface is one of the most common and popular interfaces. It acts as a cable or pipeline that can connect two networked entities. Generally, the modes of sockets are full-duplex, which allow data to be transferred bi-directionally. Here the system call select() allows the server to identify the socket that has data waiting (without polling among all the available sockets).

Clocks and Timers

In modern systems the three common types of time services are:

- To get the current time of the day.
- To get the elapsed time since a previous event.
- To set a timer to trigger event A at time T.

We all know that time operations are not standard for all the systems. To trigger the operations and to measure the elapsed time, a *Programmable Interrupt Timer*, or *PIT*, can be used. This PIT can be set to trigger an interrupt either at a specific time, or periodically on a regular basis.

- To end the time slices the scheduler uses a PIT to trigger interrupts.
- PIT can be used by the disk system for the purpose to schedule periodic maintenance cleanup.
- PIT can be used by the networks to abort or to repeat the operations that are taking a lot of time to complete.

The system clock is implemented by counting the number of interrupts generated by the PIT.

Blocking and Non-Blocking I/O

In the case of *blocking I/O*, when an I/O request is made, a process is moved to the waiting queue, and when the request gets completed, the process is moved back to the ready queue. Blocking I/O allow other processes to run in the time being.

In the case of *non-blocking I/O*, the I/O request is returned immediately, without waiting for the requested I/O operation to be completed or not. Non-blocking I/O allows the process to check the availability of the data without getting occupied completely if the data is not currently available.

Non-blocking I/O can be implemented by using the multi-threaded application, where one thread makes blocking I/O calls while other threads continue to perform other tasks. A variation of a non-blocking I/O is the *asynchronous I/O*, where the I/O request is returned immediately, allowing the process to continue with other tasks, and when the I/O operation has completed its processing and the data is made available for use, then the process is notified.

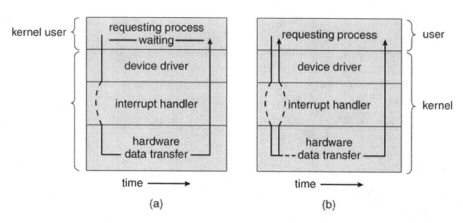

FIGURE 11.5. Two I/O methods: (a) synchronous and (b) asynchronous

⊙ Self-Quiz

Q4. Write short notes on:
  a. Block devices and character devices
  b. Network devices
  c. Clocks and timers
  d. Blocking and non-blocking I/O
Q5. Why is the Programmable Interrupt Timer used?

KERNEL I/O SUBSYSTEM

I/O Scheduling

To improve the efficiency of the system, I/O scheduling is required. If a system consists of multiple devices, then each device has a separate request queue (see Figure 11.6). Buffering and caching set the priorities, which can help in scheduling.

Buffering

The buffer or data buffer is an area of memory storage which is used to store data temporarily while moving and streaming from one place to another. Buffers can be implemented in a fixed memory location in hardware or by using a virtual data buffer in software pointing at a location in the physical memory. There are the three major reasons why buffering of I/O is performed:

1. The difference of speed between two devices. In the case of *double buffering*, two buffers are used. Here a slow device may write data into a buffer, and when that buffer gets full, the entire buffer is sent to the fast device. While the first buffer is going to the fast device, a second buffer is used by the slow device, and the two buffers alternate as each becomes full. (Double buffering is used in graphics, so that one screen image can be generated in one buffer while the other is displayed on the screen.)

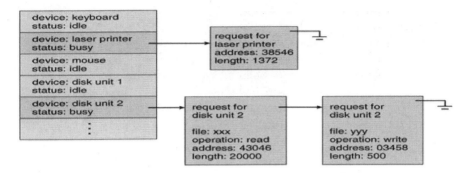

FIGURE 11.6. Device-status table

2. In networking systems buffers are used to break the messages into smaller packets to transfer from the sender side, and then to reassemble those packets on the receiving side.

3. Buffers are used to support *copy semantics*. For example, a disk write request is made by an application, and then the data is copied to the kernel buffer from the user's memory area. Now the application can change its copy of the data, but the data that ultimately gets written out to the disk is the version of the data from the time when the write request was made.

Caching

The process of keeping a *copy* of data in a faster-access location rather than the location where the data is normally stored is called *caching*.

Caching is similar to buffering. In the case of buffering, the buffer may hold only a copy of a given data item, whereas in caching a duplicate copy of some other data is stored somewhere else. Buffering and caching work in parallel, and the same storage space can be used for both purposes.

Spooling and Device Reservation

A *SPOOL (Simultaneous Peripheral Operations On Line)* buffers data for those devices that cannot support enclosed data streams, for example, printers. If at a particular time multiple processes want to be printed, then each of the processes will send their print data to files stored in the spool directory. When each file is closed, then the application is notified that the print job is completed, and finally the print scheduler will send each file to the appropriate printer one by one.

To view the spool queues, to remove jobs from the queues, to move jobs from one queue to another queue, and to change the priorities of jobs in the queue support is provided. Spool queues can be general or specific.

Error Handling

There are many reasons that I/O requests can fail. These reasons can either be temporary or permanent. To indicate the problem, I/O requests return an error bit.

I/O Protection

- All erroneous I/O (i.e., either accidental or deliberate) must be protected by the I/O system.
- All I/O requests are handled through system calls and must be performed in kernel mode rather than in user mode.
- The memory management system protects the memory-mapped areas and I/O ports.

SUMMARY

In I/O the basic hardware elements are buses, device controllers, and the devices themselves. The CPU is used as a programmed I/O or it is offloaded to the DMA controller to move data between devices and main memory. The kernel module that controls the device is called a device driver. The designing of system call interface is done to handle some basic categories of hardware, such as block devices, character devices, memory-mapped files, network sockets, and programmed interval timers. The system calls generally block the process that issues them, but the non-blocking and asynchronous calls are used by the kernel and by the applications that will not sleep while waiting for an I/O operation to be completed. Many services are provided by the kernel's I/O subsystem, such as I/O scheduling, buffering, caching, spooling, device reservation, error handling, and so on.

POINTS TO REMEMBER

- The role of the operating system is to manage and control I/O operations and I/O devices.
- To handle a particular device or category of similar devices, the kernel of an operating system is structured to use device-driver modules.
- Ports are used to connect devices with the computer.
- A bus is a set of wires and a strictly defined protocol that specifies a set of messages that can be sent over the wires.
- The four common types of buses that are used in modern PCs are: PCI, Expansion, SCSI, and Daisy Chain.
- Memory-mapped I/O is suitable for the devices that are having large data transfer quickly.
- Polling is fast and efficient.
- Non-maskable interrupts are reserved for critical error conditions.
- In interrupt chaining, each of the elements of the interrupt vector point to the head of the list of interrupt handlers.
- The procedure of handshaking between the DMA controllers and their devices takes place through two wires, DMA-request and DMA-acknowledge wires.
- A socket interface acts as a cable or pipeline that can connect two networked entities.

■ In the case of blocking I/O when an I/O request is made, a process is moved to the waiting queue, and when the request gets completed, the process is moved back to the ready queue.
■ In the case of non-blocking I/O, the I/O request is returned immediately, without waiting until the requested I/O operation has completed or not.

REVIEW QUESTIONS

Q1. When multiple interrupts from different devices appear at about the same time, a priority scheme could be used to determine the order in which the interrupts should be serviced. Discuss what issues need to be considered in assigning priorities to different interrupts.

Q2. Discuss blocking and non-blocking I/O.

Q3. Write a short note on I/O buffering.

Q4. What are the advantages and disadvantages of supporting memory-mapped I/O to device control registers?

Q5. What are the various kinds of performance overheads associated with servicing an interrupt?

Q6. Describe three circumstances in which blocking I/O should be used. Describe three circumstances in which non-blocking I/O should be used. Why not just implement non-blocking I/O and have processes busy-wait until their device is ready?

Q7. Some DMA controllers support direct virtual memory access, where the targets of I/O operations are specified as virtual addresses and a translation from virtual to physical address is performed during the DMA. How does this design complicate the design of the DMA controller? What are the advantages of providing such a functionality?

CHAPTER 12

DISK MANAGEMENT

Purpose of the Chapter

Bulk data is stored in external devices such as hard disks, magnetic drives, and so on. Disks are the secondary storage media. Disk management, which includes disk scheduling, disk buffering, swap space management, and so on, is done by the operating system. In this chapter, we discuss various disk scheduling techniques to fulfill the input/output request to access the disk. Further in the chapter, the concept of swap space management is explained, which provides extra space for the process to grow without deferring the execution of other processes.

INTRODUCTION

Many types of storage devices were in use in the early days, for example, magnetic disks, magnetic tapes, floppy disks, and so on. These devices are no longer in use and have been replaced by hard disks that are high in capacity as well as access speed. Disks help in storing bulk amounts of data in computer systems.

STRUCTURE OF THE DISK

Modern disk drives are addressed as large one-dimensional arrays (or linear arrays) of logical blocks, where the logical block is the smallest unit of transfer. Usually the size of a logical block is 512 bytes, but some of the disks can be formatted as low-level to have a different logical block size, such as 1,024 bytes. The one-dimensional array of logical blocks is sequentially mapped onto the sectors of the disk. Hence, sector 0 is the first sector of the first track on the outermost cylinder. The order in which the mapping proceeds is through that track, then through the rest of the tracks in that cylinder, and then through the rest of the cylinders from outermost to innermost.

This mapping technique theoretically converts a logical block number into an old-style disk address that is composed of a cylinder number, a track number within that cylinder, and a sector number within that track. But practically this conversion is difficult to perform because of the following two reasons:

1. Most disks have some defective sectors, but the mapping hides this by substituting spare sectors from elsewhere on the disk.
2. The number of sectors per track is not a constant on some drives.

Let's elaborate on the second reason. On the media that use Constant Linear Velocity (CLV), the density of bits per track is uniform. The further a track is from the center of the disk, the greater its length, and therefore it can hold more sectors. The number of sectors per track decreases as we move from outer zones to inner zones. In the outermost zone tracks typically hold 40% more sectors than the tracks in the innermost zone. The drive increases its rotation speed as the head moves from the outer to the inner tracks to keep the same rate of data moving under the head, as in CD-ROM and DVD-ROM drives, for example. On the other hand, the disk rotation speed can stay constant, and the density of bits decreases from inner tracks to outer tracks to keep the data rate constant. This method is used in hard disks and is known as Constant Angular Velocity (CAV).

The number of sectors per track has been increasing as disk technology improves, and the outer zone of a disk usually has several hundred sectors per track. Similarly, the number of cylinders per disk has been increasing.

DISK SCHEDULING

The processing speed of the CPU (Central Processing Unit) and access speed in the main memory is much higher than the access speed of a hard disk. To cope with the speed of the processor and main memory, it is necessary to schedule the request of I/O (Input/Output). *Scheduling the I/O requests on hard disk is called disk scheduling*. It helps in decreasing the access time from the disk and increases overall throughput of the system. *The total time to fetch some data from the disk is called access time*. This time should always be small for high throughput of the computer system. Access time has two major components, which are *seek time* and *rotational latency*.

- **Seek Time:** The time required for the disk arm to move the heads to the cylinder, which contains the required sector (the place in the disk where required data is stored), is called seek time.
- **Rotational Latency:** Time taken to rotate the disk arm to the desired sector in the given track is called rotational latency. It is also known as rotational delay.

The sum of seek time and rotational latency is called access time of disk. Hence access time can also be defined as the total time to fetch a block from the disk. As explained in earlier chapters, many processes are concurrently executed in the computer system. Hence several I/O requests are generated by the processes. The OS maintains a queue of all the requests from processes. This queue of I/O requests needs to be scheduled so that total access time reduces and system efficiency increases. For this purpose various scheduling algorithms are discussed further. The scheduling algorithms are as follows:

- First Come First Serve (FCFS) Scheduling
- Shortest Seek Time First (SSTF) Scheduling
- SCAN Scheduling
- C-SCAN Scheduling
- LOOK Scheduling
- C-LOOK Scheduling

First Come First Serve (FCFS) Scheduling

In this scheduling algorithm, the request which comes first will be served first. This method is considered as the simplest method of scheduling. It processes the request in sequence order. Requests are processed in the order in which the input/output requests arrive.

Example 12.1: Suppose that a disk drive has 200 cylinders, numbered from 0 to 199. The queue which is maintained is 86, 172, 28, 132, 13, 150, 55, 60, 20, 180 in order. The disk arm is initially at cylinder 40. Starting from the current head position, what is the total distance (in cylinders) that the disk arm moves to satisfy all the pending requests for FCFS scheduling algorithm?

Solution: The disk arm will move from 40 to 86 then 86 to 172, then 172 to 28, then 28 to 132, and so on to 13, 150, 55, 60, 20, 180. The working of the FCFS algorithm for the example given is shown in Figure 12.1.

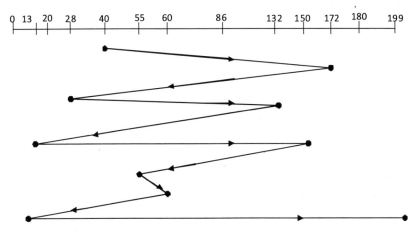

FIGURE 12.1. FCFS disk scheduling

Total head movements = $(86 - 40) + (172 - 86) + (172 - 28) + (132 - 28)$
$+ (132 - 13) + (150 - 13) + (150 - 55) + (60 - 55) + (60 - 20) + (180 - 20)$
$= 46 + 86 + 144 + 104 + 119 + 137 + 95 + 5 + 40$
$+ 160$ head movements
$= 936$ head movements

Drawback: In this method, the seek time is high because the disk arm has to cover the long distance to fulfill the request in order. As shown in the example, the movement of arm from 172 to 28 to 132 to 15 increases the mean seek time.

Shortest Seek Time First (SSTF) Scheduling

In this method the disk arm selects those I/O operations that require minimum tracks to be traversed. The request close to the disk head is served first. The SSTF method helps in achieving the minimum seek time serving the request which requires the minimum seek time. This method gives better performance than FIFO.

Example 12.2: Suppose that a disk drive has 200 cylinders, numbered from 0 to 199. The queue which is maintained is 86, 172, 28, 132, 13, 150, 55, 60, 20, 180 in order. The disk arm is initially at cylinder 40. Starting from the current head position, what is the total distance (in cylinders) that the disk arm moves to satisfy all the pending requests for the SSTF scheduling algorithm?

Solution: The disk arm will move from 40 to 28, then 28 to 20, then 20 to 13, and so on. The working of the SSTF algorithm for the example given is shown in Figure 12.2.

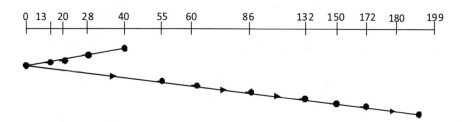

FIGURE 12.2. SSTF disk scheduling

Total head movements = $(40 - 28) + (28 - 20) + (20 - 13) + (40 - 13) + (55 - 40) +$
$(60 - 55) + (86 - 60) + (132 - 86) + (150 - 132) + (172 - 150) + (180 - 172)$
$= 12 + 8 + 7 + 27 + 15 + 5 + 26 + 46 + 18 + 22$
$+ 8$ head movements
$= 194$ head movements

Drawback: This method may suffer from starvation. When requests near to the disk head arrive frequently, then the long-distance request may suffer from starvation.

SCAN Scheduling

In this method, the disk head is required to move to the beginning of the disk and from there towards the end of the disk by satisfying the request which comes in the root of the disk head. This algorithm is also known as the elevator algorithm, because the way it works is similar to the elevator of the building which covers all the floors going up and then down, leaving people on their desired floor accordingly. This method behaves like SSTF, because it starts moving from the lower cylinder number.

Example 12.3: Suppose that a disk drive has 200 cylinders, numbered from 0 to 199. The queue which is maintained is 86, 172, 28, 132, 13, 150, 55, 60, 20, 180 in order. The disk arm is initially at cylinder 40. Starting from current head position, what is the total distance (in cylinders) that the disk arm moves to satisfy all the pending requests for the SCAN scheduling algorithm?

Solution: The disk arm will move from 40 to 28, then 28 to 20, then 20 to 13, then 13 to 0, and so on. The working of the SCAN algorithm for the example given is shown in Figure 12.3.

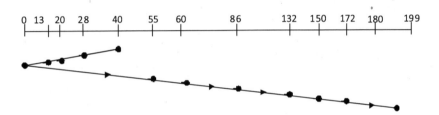

FIGURE 12.3. SCAN disk scheduling

Total head movements = $(40 - 28) + (28 - 20) + (20 - 13) + (13 - 0) + (55 - 0) + (60 - 55) + (86 - 60) + (132 - 86) + (150 - 132) + (172 - 150) + (180 - 172)$
$$= 12 + 8 + 7 + 13 + 55 + 5 + 26 + 46 + 18 + 22$$
$$+ 8 \text{ head movements}$$
$$= 220 \text{ head movements}$$

Drawback: In this algorithm, the disk head starts from the beginning no matter what other requests are present on the other end.

C-SCAN Scheduling

This scheduling algorithm is an improvement of SCAN scheduling. In this method, the disk starts from the end where more requests are present. While going to the end it satisfies the entire request which comes under the roof. When the disk arm reaches the arm, it quickly returns to the other end without fulfilling any request on the way.

Example 12.4: Suppose that a disk drive has 200 cylinders, numbered from 0 to 199. The queue which is maintained is 86, 172, 28, 132, 13, 150, 55, 60, 20, 180, in order. The disk arm is initially at cylinder 40. Starting from current head position, what is the total distance (in cylinders) that the disk arm moves to satisfy all the pending requests for the C-SCAN scheduling algorithm?

Solution 12.4 The disk arm will move from 40 to 55, then 55 to 60, then 60 to 86, and so on. The working of the C-SCAN algorithm for the example given is shown in Figure 12.4.

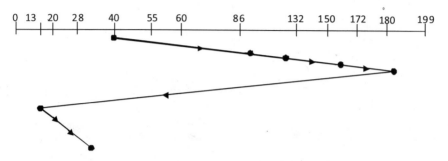

FIGURE 12.4. C-SCAN disk scheduling

Total head movements = $(55 - 40) + (60 - 55) + (86 - 60) + (132 - 86) + (150 - 132) + (172 - 150) + (180 - 172) + (199 - 180) + (199 - 0) + (13 - 0) + (20 - 13) + (28 - 20)$

$$= 15 + 5 + 26 + 46 + 18 + 22 + 8 + 19 + 199 + 13 + 7 + 8 \text{ head movements}$$
$$= 386 \text{ head movements}$$

Drawback: The drawback of this method is that it services the request only in one direction.

Note: *The SCAN and C-SCAN scheduling algorithms suffer from one major drawback: the disk head moves to the end of the disk no matter whether the request is present or not. This increases the sectors which are traversed by the disk arm unnecessarily. The refinement of this algorithm is done in the LOOK and C-LOOK algorithms.*

LOOK Scheduling

In this method the head looks for a request before continuing in the given direction. It is similar to the SCAN algorithm. The arm will go to the final request in each direction. It reverses the direction immediately without going all the way to the end of the disk.

Example 12.5: Suppose that a disk drive has 200 cylinders, numbered from 0 to 199. The queue which is maintained is 86, 172, 28, 132, 13, 150, 55, 60, 20, 180, in order. The disk arm is initially at cylinder 40. Starting from current head position, what is the total distance (in cylinders) that the disk arm moves to satisfy all the pending requests for the LOOK scheduling algorithm?

Solution: The disk arm will move from 40 to 28, then 28 to 20, then 20 to 13, and so on. The working of the LOOK algorithm for the example given is shown in Figure 12.5.

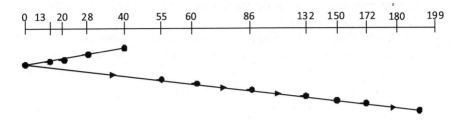

FIGURE 12.5. LOOK disk scheduling

Total head movements = (40 – 28) + (28 – 20) + (20 – 13) + (40 – 13) + (55 – 40) + (60 – 55) + (86 – 60) + (132 – 86) + (150 – 132) + (172 – 150) + (180 – 172)

$$= 12 + 8 + 7 + 27 + 15 + 5 + 26 + 46 + 18 + 22$$
$$+ \ 8 \ \text{head movements}$$
$$= 194 \ \text{head movements}$$

C-LOOK Scheduling

In this method the head looks for a request before continuing in the given direction. It is similar to the C-SCAN algorithm. Here C-LOOK stands for circular LOOK. The arm will go to the final request in each direction. It reverses the direction immediately without going all the way to the end of the disk.

Example 12.5: Suppose that a disk drive has 200 cylinders, numbered from 0 to 199. The queue which is maintained is 86, 172, 28, 132, 13, 150, 55, 60, 20, 180, in order. The disk arm is initially at cylinder 40. Starting from the current head position, what is the total distance (in cylinders) that the disk arm moves to satisfy all the pending requests for the C-LOOK scheduling algorithm?

Solution: The disk arm will move from 40 to 55, then 55 to 60, then 60 to 86, and so on. The working of the C-LOOK algorithm for the example given is shown in Figure 12.6.

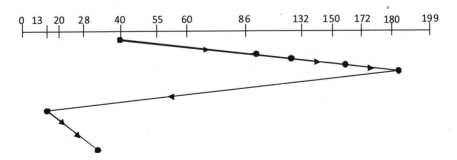

FIGURE 12.6. C-LOOK disk scheduling

Total head movements = (55 – 40) + (60 – 55) + (86 – 60) + (132 – 86) +
(150 – 132) + (172 – 150) + (180 – 172) + (180 – 13) + (20 – 13) + (28 – 20)
$$= 15 + 5 + 26 + 46 + 18 + 22 + 8 + 167 + 7$$
$$+ 8 \text{ head movements}$$
$$= 322 \text{ head movements}$$

(•) Self-Quiz

Q1. Explain FCFS disk scheduling with an example.
Q2. Discuss SSTF disk scheduling with an example.
Q3. Describe SCAN disk scheduling with an example.
Q4. Illustrate C-FCFS disk scheduling with an example.
Q5. Explain LOOK disk scheduling with an example.
Q6. Describe C-LOOK disk scheduling with an example.

SECONDARY STORAGE MANAGEMENT

Formatting of the Disk

A new magnetic disk is considered as a plate of magnetic recording material. Before storing data on a disk, it should be divided into sectors that can be read or written by the disk controller. This process is known as low-level formatting (or physical formatting), which fills each sector of the disk with a special data structure. This data structure typically contains a header, a data area (generally of 512 bytes), and a trailer. The header and trailer contain information (such as a sector number and an Error-Correcting Code (ECC)) that is used by the disk controller. During normal I/O, when the controller writes a sector of data, the ECC will be updated by a value calculated from all the bytes in the data area. When the controller is reading the sector, then the ECC is recalculated and it is compared with the stored value. If the stored value does not match the calculated value, then it indicates that the data area of the sector has become corrupted and the disk sector may be bad. The ECC processing is done automatically whenever a sector is read or written.

To make use of a disk for holding files, the operating system records its own data structures on the disk. The process consists of two steps:

1. To partition the disk into one or more groups of cylinders. Each partition is thought of as a separate disk. For example, one partition can hold a copy of the operating system's executable code, while the other holds the user files.
2. To perform logical formatting, in which the operating system stores the initial file system data structures onto the disk. These data structures may include maps of free and allocated space and an initial empty directory.

Most of the file systems group the blocks together to form larger chunks (known as clusters), which results in increased efficiency. Blocks are used for disk I/O whereas the file system I/O is done through clusters. A special program is provided by some of the operating systems to use a disk partition as a large sequential array of logical blocks, without any file system data structures. This array is also known as the raw disk, and the I/O for this array is known as raw I/O.

Booting from the Disk

To start a computer (i.e., fresh start or reboot), there are some initial programs to run (Bootstrap). The bootstrap program initializes all aspects of the system, such as CPU registers, device controllers, and the contents of main memory, and starts the operating system. For its processing, the bootstrap program finds the operating system kernel on the disk, loads that kernel into memory, and jumps to an initial address to begin the operating system execution.

For most computers, the bootstrap is stored in Read Only Memory (ROM), because no initializing is required by ROM and the location is fixed so that the processor can start executing when the system is powered up or reset. And also because ROM is read only, a computer virus cannot infect it. To change the bootstrap code it is required to change the ROM (hardware chips), which is a problem. Due to this reason only, most of the systems store a tiny bootstrap loader program in the boot ROM whose only job is to bring a full bootstrap program from disk. The changes can be made easily in a full bootstrap program. The full bootstrap program is stored in "the boot blocks" at a fixed location on the disk. A disk that has a boot partition is called a boot disk or system disk.

Management of Bad Blocks

We know that disks contain moving parts and have small tolerance to disturbance, and therefore there is a possibility of failure. Most of the disks come with bad blocks from the factory. These blocks can be handled through various methods depending on the disk and the controller.

Simple disks, that is, disks with IDE (Integrated Development Environment) controllers, handle bad blocks manually. For example, the MS-DOS (Microsoft Disk Operating System) format command performs logical formatting and scans the disk to find bad blocks. If a bad block is found, then it writes a special value into the corresponding FAT (File Allocation) entry, which tells the allocation routines to not to use that block. On the contrary, if a bad block is found during normal operation, then a special program, *chkdsk*, must run manually to search for the bad blocks and lock them. Generally the data that exist on the bad blocks are lost.

In sophisticated disks, such as the SCSI disks, the controller maintains a list of bad blocks on the disk. The initialization of this list is done during the low-level formatting at the factory and is updated over the life of the disk.

Low-level formatting also sets a spare sector that is not visible to the operating system. The controller can be instructed to replace the bad sector logically with one of the spare sectors. This scheme is known as *sector sparing* or forwarding.

On the other hand, some controllers are instructed to replace a bad block by sector slipping, rather than by sector sparing.

⊙ Self-Quiz

Q7. Explain disk formatting.
Q8. What do you understand about bad blocks?
Q9. Explain the concept of booting from the disk.

SWAP-SPACE MANAGEMENT

For an operating system, swap-space management is a low-level task. Disk space is used by virtual memory as an extension of the main memory. As we know that accessing the disk is slower than accessing the memory, the use of swap space will decrease the system performance. The main aim for designing and implementing the swap space is to provide the best throughput for the virtual memory system.

What Is Swap-Space?

Depending on the memory management algorithms, swap-space can be used differently for different operating systems. For example, the systems that implement swapping may use swap space to hold the entire process image, which includes the code and data segments. Whereas the system that implements paging will simply store the pages that have been pushed out of the main memory. The amount of swap-space required by a system may vary. This variation depends upon the amount of physical memory, the amount of virtual memory, and the method in which the virtual memory is used.

It is good to overestimate the amount of required swap-space rather than to underestimate it. The reason for this is if a system runs out of swap-space, then it may be forced to abort processes or may crash completely. The only drawback of overestimation is that it wastes the disk space that can be used by other files.

Location of Swap-Space

A swap-space can be located in one of two places; either it can be engraved out of the normal file system or it can be in a separate disk partition. If the swap-space is a large file within the file system, then its creation, naming, and allocation of space can be done through normal file-system routines. Although this approach is easy to implement, it is not efficient. External fragmentation increases the swapping times by making multiple seeks during reading or writing of a process image.

On the other hand, swap-space can be created in a separate raw partition, where no file system or directory structure is placed. To allocate and to deallocate the blocks from the raw partition, a separate swap-space storage manager is used. This manager uses those algorithms optimized for speed rather than for storage, because swap space can be accessed more frequently than the file systems. Internal fragmentation may be increased because generally the life of data in the swap-space is shorter than that of the files in the file system. Some operating systems are flexible and can swap both in raw partitions and in file-system space.

Example of Swap-Space Management

Here it is demonstrated how swap-space is used by following the evolution of swapping and paging in various UNIX systems. Initially swapping was implemented with the traditional UNIX kernel, which copies the entire processes between the contiguous disk regions and memory. Later UNIX was developed with a combination of swapping and paging. The standard UNIX methods were changed in Solaris 1 (SunOS) by the designers. These changes were made to improve the efficiency and to reflect technological changes. More changes were made in the later versions of Solaris. The major change made in Solaris was that it can now allocate swap-space only when a page is forced out of physical memory, instead of when the virtual memory page is first created.

Linux is similar to Solaris, where the swap space is only used for unspecified memory or for the regions of memory that can be shared by several processes.

⊙ Self-Quiz

Q10. What is swap-space? Give an example of swap-space management.
Q11. Write the drawback of overestimation.

RAID

RAID stands for Redundant Array of Independent Disks. It consists of seven levels, level 0 through level 6. The design architecture of RAID shows the following characteristics:

1. RAID is a set of physical disk drives viewed by the operating system as a single logical drive.
2. Data are distributed across the physical drives of an array. This is known as striping.
3. In the case of a disk failure, redundant disk capacity is used to store parity information, which guarantees the recovery of data.

Levels of RAID

Table 12.1 roughly describes all seven levels of RAID. The Table shows the I/O performance both in the terms of data transfer capacity (or ability to move data), and I/O request rate (or ability to satisfy I/O requests). The four commonly used RAID levels are 0, 1, 5, and 6.

Table 12.1. Levels of RAID and Their Description

Category	Level	Description	Disks required	Data availability	Large I/O data transfer capacity	Small I/O request rate
Striping	0	Nonredundant	N	Lower than single disk	Very high	Very high for both read and write
Mirroring	1	Mirrored	$2N$	Higher than RAID 2,3,4, or 5; lower than RAID 6	Higher than single disk for read; similar to single disk for write	Up to twice that of a single disk for read; similar to single disk for write
Parallel access	2	Redundant via Hamming code	$N + m$	Much higher than single disk; comparable to RAID 3,4, or 5	Highest of all listed alternatives	Approximately twice that of a single disk
	3	Bit-interleaved parity	$N + 1$	Much higher than single disk; comparable to RAID 2, 4, or 5	Highest of all listed alternatives	Approximately twice that of a single disk
	4	Block-interleaved parity	$N + 1$	Much higher than single disk; comparable to RAID 2, 3, or 5	Similar to RAID 0 for read; significantly lower than single disk for write	Similar to RAID 0 for read; significantly lower than single disk for write
Independent access	5	Block-interleaved distributed parity	$N + 1$	Much higher than single disk; comparable to RAID 2, 3, or 4	Similar to RAID 0 for read; lower than single disk for write	Similar to RAID 0 for read; generally lower than single disk for write
	6	Block-interleaved dual distributed parity	$N + 2$	Highest of all listed alternatives	Similar to RAID 0 for read; lower than RAID 5 for write	Similar to RAID 0 for read; significantly lower than RAID 5 for write

N = number of data disks; m proportional to log N

RAID Level 0

To improve the performance and to provide data protection, level 0 does not include redundancy. Therefore, RAID level 0 is not considered as a true member of the RAID family. However, there are some applications where performance and capacity are prime concerns and low cost is more important than improved reliability.

In level 0, the user data and system data are distributed through all of the disks in the array. The advantage of this is that if there are two different I/O requests pending for two different blocks of data, then there is a possibility that the requested blocks are on different disks. Hence, the two requests can be issued in parallel; this reduces the I/O queuing time. In level 0, the data are striped across all the available disks. Figure 12.7 depicts the RAID level 0, where the entire user data and system data are viewed as being stored on a logical disk. This logical disk is divided into strips (these strips may be physical blocks, sectors, or some other unit). The advantage of this design is that if a single I/O request constitutes multiple logically contiguous strips, then up to n number of strips for that request can be handled in parallel, which reduces the I/O transfer time.

FIGURE 12.7. RAID level 0

RAID Level 1

The method to calculate redundancy of RAID level 1 is different from the RAID levels 2 through 6. In RAID level 1 redundancy is calculated by duplicating all the data, whereas in other RAID levels, redundancy is determined by calculating some form of parity. As shown in Figure 12.8, RAID 1 uses data striping as it is used in RAID 0, but the difference is that in RAID 1, each logical strip is mapped to two distinct physical disks, so that every disk in the array has a mirror disk that contains the same data. RAID 1 can be implemented without data striping (not a common method). The main features of RAID 1 organization are as follows:

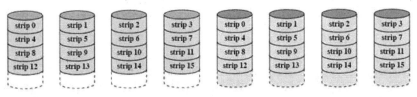

(b) RAID 1 (mirrored)

FIGURE 12.8. RAID level 1 (mirrored)

1. To service a read request, any of the two disks that hold the requested data and involve the minimum seek time plus rotational latency can be used.
2. To service a write request, it is required that both the strips be updated in parallel.
3. Recovery is simple. When one drive fails, then the data may be accessed from the second drive.

The major disadvantage of RAID 1 is its cost, as it uses two logical disks, hence twice the disk space that is required by RAID 1.

RAID Level 2

RAID level 2 and RAID level 3 use the technique of parallel access, in which all the members of disk participate in the execution of each I/O request.

RAID level 2 also uses the concept of data striping. The strips of RAID 2 and RAID 3 are very small (as a single byte or word). For RAID 2 usually a Hamming code is used, through which a single-bit error can be corrected and double-bit errors can be detected as shown in Figure 12.9. Though the number of disks used by RAID 2 is less than RAID 1, it is also costly. RAID 2 can be used in those environments that have many disk errors.

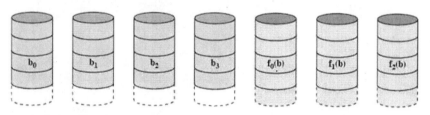

(c) RAID 2 (redundancy through Hamming code)

FIGURE 12.9. RAID level 2 (redundancy through hamming code)

RAID Level 3

The organization of RAID 3 is similar to that of RAID 2. The only difference is that a single redundant disk is required by RAID 3. Parallel access is employed by RAID 3. Rather than using the error-correcting code, a simple parity bit is computed for the set of individual bits in the same position on all of the data disks.

In the case of a drive failure, the parity drive is accessed and data is restored from the remaining devices. When the failed drive is replaced, then the missing data can be restored on new drive and operation resumed. Restoring of data is simple.

In the case of a disk failure, all of the data is available in reduced mode. In this mode, for read operations, the missing data are reinforced by calculating exclusive-OR. When a write operation is performed to a reduced RAID 3 array, consistency of the parity must be maintained to regenerate it (See Figure 12.10).

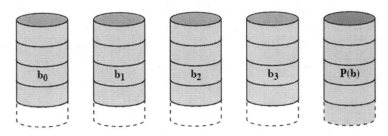

(d) RAID 3 (bit-interleaved parity)

FIGURE 12.10. RAID level 3 (bit-interleaved parity)

As the striping of data is done into very small strips, RAID 3 can achieve very high data transfer rates.

RAID Level 4

RAID levels 4, 5, and 6 use an independent access technique, where each member of the disk operates independently, so that the separate I/O requests can be satisfied in parallel.

RAID level 4 also uses data striping (data strips are large in the RAID levels 4, 5, and 6). In RAID 4, a bit-by-bit parity strip is calculated across the corresponding strips on each of the data disks, and the parity bits are stored in the corresponding strip on the parity disk.

When the small size of I/O request is performed, then RAID 4 involves a write penalty. Whenever the write operation occurs, the array management software must update the user data as well as the corresponding parity bits.

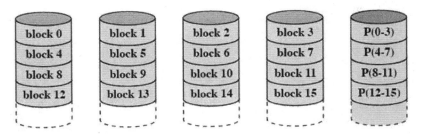

(e) RAID 4 (block-level parity)

FIGURE 12.11. RAID level 4 (block-level parity)

RAID Level 5

The organization of RAID level 5 is similar to RAID level 4. The only difference is that RAID level 5 distributes the parity strips across all the disks. The typical allocation is a Round-Robin scheme, which is represented in Figure 12.12. Consider an n-disk array: for the first n strips the parity strip is on a different disk, and then the pattern is repeated.

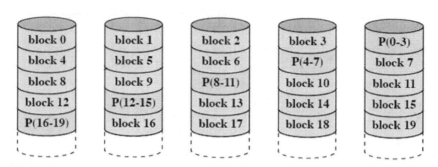

(f) RAID 5 (block-level distributed parity)

FIGURE 12.12. RAID level 5 (block-level distributed parity)

RAID Level 6

Two different parity calculations are carried out in the RAID level 6, and these parities are stored in separate blocks on different disks. Therefore, a RAID 6 array whose user data requires N disks consists of $N + 2$ disks.

Figure 12.13 explains the scheme. Here P and Q are two different data check algorithms. One is the exclusive-OR calculation used in RAID level 4 and level 5. The other is an independent data check algorithm. Because of this, it is possible to regenerate data even if two disks containing user data fail.

The main advantage of RAID 6 is that it provides tremendously high data availability.

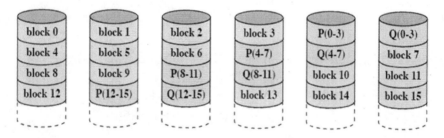

(g) RAID 6 (dual redundancy)

FIGURE 12.13. RAID level 6 (dual redundancy)

 Self-Quiz

Q12. What is RAID?
Q13. Explain all the levels of RAID.

SUMMARY

In this chapter, you have learned about disk management. Using this, you can perform various administrative tasks, like partitioning layouts or volumes, initializing or upgrading disks, and choosing protection strategies, such as mirroring and striping. You have also learned about disk scheduling. Servicing the disk I/O requests is known as disk scheduling. To use hardware efficiently, disk scheduling is required, which includes fast access time using seek time and rotational latency, and large disk bandwidth.

Effective bandwidth, the average response time, and the discrepancy in response time can be improved with disk scheduling algorithms such as SSTF, SCAN, C-SCAN, LOOK, and C-LOOK. These improved algorithms are made through strategies for disk queue ordering.

Disk blocks are managed by the operating system. First of all a disk must be formatted to create the sectors on the raw hardware (a new disk is usually preformatted). Then partitioning takes place on the disk after which the file systems are created, and the boot blocks are allocated to store the bootstrap program. When a corrupted block is encountered, the system must either lock out that block or replace that block logically with a spare block.

You have also learned about the basics of a redundant array of inexpensive disks. RAID is a data storage virtualization technology that combines multiple disk drive components into a logical unit for the purposes of data redundancy and performance improvement.

To have an efficient swap-space results in good performance. Systems generally use raw disk access for paging I/O rather than using the file system. For large systems, disks are frequently made redundant via RAID algorithms. With these algorithms more than one disk can be used for a given operation. In the case of failure it continues its operation and provides automatic recovery. The different levels of RAID algorithm provide some combination of reliability and high transfer rates.

POINTS TO REMEMBER

- The disk helps in storing a bulk amount of data in computer systems.
- Scheduling the I/O requests on a hard disk is called disk scheduling.
- The total time to fetch some data from a disk is called access time.
- In the FCFS algorithm the first request will be served first. It suffers from high seek time.
- The seek time of the SSTF algorithm is low, but it may suffer from starvation.
- In the SCAN algorithm, the disk head starts from the beginning.
- The C-SCAN algorithm services the requests only in one direction.
- The refinement of SCAN and C-SCAN is done in the LOOK and C-LOOK algorithms.
- Formatting a disk with a larger sector size means that fewer sectors can fit on each track. It also means that fewer headers and trailers are written on each track and more space is available for user data.
- To start a computer there is an initial program to run. This initial program is Bootstrap.

- The improvement in the performance is done by caching the block location information in physical memory and by using a special tool to allocate physically contiguous blocks for the swap file, but the cost of traversing the file-system data structures still remains.
- Recovery is simple. When one drive fails, then the data may be accessed from the second drive.

REVIEW QUESTIONS

Q1. Write down the criteria for the selection of a disk-scheduling algorithm.

Q2. Suppose a disk has 5000 cylinders, number 0 to 4999. The drive is currently sending a request cylinder 143 and the previous was cylinder 125. The queue of the pending request in FIFO order 86, 1470, 913, 1774, 948, 1509, 1022, 1750, 130.

Q3. Discuss the disk scheduling algorithm.

Q4. What are the essential goals of disk scheduling? Why is each im portant?

Q5. Compare the throughput of C-SCAN with that of SCAN.

Q6. Explain the SSTF and SCAN disk scheduling policies. Obtain the total number of head movements needed to satisfy the following sequence of track requests for each of the two policies: 27, 129, 110, 186, 147, 41, 10, 64, 120

Q7. Assume that the disk head is initially positioned over track 100 and is moving in the direction of decreasing track number.

Q8. Write short notes on:
   a. Disk scheduling policies
   b. RAID

Q9. Suppose the moving head disk with 200 tracks is currently serving a request for 143 and just finished a request for track 125. If the request queue is kept in FIFO order 86, 147, 91, 177, 94, 150, calculate the total head movement for the following scheduling:
   a. FCFS
   b. SSTF
   c. C-SCAN

CHAPTER 13

CHARACTERIZATION OF A DISTRIBUTED SYSTEM

Purpose of the Chapter

The chapter gives the proper definition of a distributed system. Some of the significant consequences are defined for a better understanding of the distributed system. A comparative study of the distributed system with a parallel system and centralized system is done. This chapter also focuses on the design issues of a distributed system and examples of distributed systems.

INTRODUCTION

A distributed system is composed of several computers that do not share a common memory or a clock. These computers communicate with each other by passing messages over a communication network, and each of these computers run their own operating system and have their own memory. Some very famous and important definitions are given below.

According to Leslie Lamport, "A distributed system is one that stops you from getting any work done whenever a machine you have never even heard of crashes."

According to Brazier, "A distributed system is a collection of autonomous computers linked by a network, with software designed to produce an integrated computing facility."

According to Tanenbaum, "A distributed system is a collection of independent computers that appear to the users of the system as a single computer."

CONSEQUENCES OF A DISTRIBUTED SYSTEM

Some of the significant consequences for understanding distributed systems are as follows:

■ **Concurrency:** The convention for any network of computers is its concurrent program execution, which says that the user "x" will perform work on his or her computer and the user "y" will perform work on his or her computer. They are allowed to share resources (such as Web and files) when required. By the addition of more resources (such as computers) to the network, the system capacity will be increased for handling shared resources.

■ **No global clock:** The only way to hold communication in a distributed system is by sending messages through the network. It is not possible to have synchronization among multiple computers and does not guarantee to have synchronization over time; therefore, events are ordered logically. It is not possible to have a process that could be aware of a single global state.

■ **Independent failures:** Whenever faults occur in the network, it results in the isolation of computers and those who are connected to them. But isolating those computers does not mean that they will stop running; this is due to the fact that programs on those computers may not be able to detect whether the network has failed or it has become unusually slow. In the same manner, unexpected termination of a program or crash (failure) of a computer does not broadcast immediately to the other components with which they communicate.

When a group of users are working together, they need to communicate with each other, to share data and resources. The requirements of distributed systems came from sharing resources among users. These resources could be hardware (such as disks and printers) or software (such as files, databases, and data objects of all kinds).

CENTRALIZED SYSTEM VERSUS DISTRIBUTED SYSTEM

The following table summarizes the comparison between a centralized system and a distributed system.

TABLE 13.1. Comparison between a Centralized System and a Distributed System

Centralized System	Distributed System
Has non-autonomous components.	Has autonomous components.
This type of system has been often built by using homogeneous technology.	These types of systems may be built by using heterogeneous technology.
Here resources of a centralized system are shared among multiple users all the time.	Distributed system components may be used exclusively and executed in concurrent process.
Have single point of control and failure.	Have multi point failure.

PARALLEL SYSTEMS VERSUS DISTRIBUTED SYSTEMS

The following table summarizes the comparison between a parallel system and a distributed system.

TABLE 13.2. Comparison between a Parallel and a Distributed System

Parameter	Parallel System	Distributed System
Memory	Tightly coupled shared memory. For example, UMA (Unified Memory Architecture) and NUMA (Non-Uniform Memory Access).	Message passing, RPC (Remote Procedure Call) and/or use of distributed shared memory.
Control	Global clock control is needed. For example, SIMD (Single Instruction Multiple Data), MIMD (Multiple Instruction Multiple Data).	No global clock control is required. Synchronization algorithm is required.
Processor Interconnection	Bus, Mesh, Tree, Hypercube network, etc.	(Bus) Ethernet, token ring, and SCI (Serial Communication Interface) rings, switching network, etc.
Main Focus	Performance computing.	Performance, Reliability/availability, Information/resource sharing.

Self-Quiz

Q1. What is a distributed system? Why it is required?

Q2. Compare the distributed system and the centralized system.

Q3. Why is the distributed system better than the parallel system? Explain.

RESOURCE SHARING AND THE WEB

In a distributed system, resources are:

- Encapsulated within computers.
- They can be accessed from other computers only by communication.
- Can be managed by a program that offers a communication interface.

Resources may be shared in either form (e.g., printer, scanner, etc.) to reduce costs. But the concern of users for sharing resources is to share the higher-level resources that play a part in their application and in their everyday work and social activities.

An example is the sharing of data in the form of a shared database or a set of Web pages. According to users, shared resources can be a search engine or a currency converter, without regard for the servers that provide these.

In practice, the patterns of resource sharing vary widely depending upon their scope and on how closely users work together. At one extreme, a search engine on the Web provides a facility to the users throughout the world (users who never come in contact with each other directly). At the other extreme, in Computer Supported Cooperative Working (CSCW), the group of users (who cooperate directly) share resources such as documents in a small closed group. To coordinate users' actions, what mechanism must be supplied by the system is determined by the pattern of sharing and the geographic distribution of particular users.

Service is the term used for a distinct part of a computer system that manages a collection of related resources and presents their functionality to the users and application through a well-defined set of operations.

Server refers to a running program on a networked computer which accepts request messages from programs that are running on the other computers (i.e., client) to perform a service and to provide an appropriate response by replying to messages. A complete interaction between client and server is called a *remote invocation*.

Many of the distributed systems can be constructed completely in the form of interacting clients and services, for example, the World Wide Web, e-mail, networked printers, and so on. An executing Web browser is an example of a client. The Web browser communicates with the Web servers to request Web pages from the server.

Self-Quiz

Q1. Explain the following terms:
   a. **WWW**
   b. **Client**
   c. **Server**
   d. **Remote invocation**
   e. **Service**
Q2. Name any six types of hardware or software resources that can be shared efficiently. Also give examples of their sharing as it occurs in a distributed system.

DESIGN ISSUES OF A DISTRIBUTED OPERATING SYSTEM

Various challenges to build a distributed system are:

1. **Heterogeneity:** The Internet enables the users to access services and to run applications over a heterogeneous collection of computers and networks. Heterogeneity is applied on the following:

   • **Computer hardware:** For example, Mainframe, workstation, PCs (Personal Computers), services, etc.
   • **Computer software:** For example, UNIX, MS (Microsoft) Windows, International Business Machine/ Operating System 2 (IBM OS/2) or Real-time Operating Systems, etc.
   • **Unconventional devices:** For example, teller machines, telephone switches, robots, manufacturing systems, etc.
   • **Diverse Networks and Protocols:** For example, Ethernet, FDDI (Fiber Distributed Data Interface), ATM (Automated Teller

Machine), TCP/IP (Transmission Control Protocol/Internet Protocol), Novell Netware, etc.

The information that all the connected computers are communicating to each other using Internet protocols is masked for different networks. Some approaches to do so are:

- Middleware (e.g., CORBA or Common Object Request Broker Architecture)
- Mobile code (e.g., Java applet)

Middleware is applied to the software layer, which provides a programming abstraction as well as masking the heterogeneity of the underlying network, hardware, operating systems, and programming languages.

Mobile code is code that can be sent from one computer to the other computer and run at the destination.

2. **Openness:** The openness is the characteristic of a computer system which determines that whether the system can be extended or re-implemented in various ways. The openness of a distributed system can primarily be determined by the degree to which new resource sharing services can be added, and can be made available for use by various client programs.

The step to provide openness is to publish the interfaces of the components, but to integrate these components that are written by different programmers is a real challenge. Systems designed to support resource sharing in this way is termed as *open distributed systems*, for emphasizing the fact that they are extensible.

The open distributed systems:

- Can be characterized by the fact that their key interfaces are published.
- Are based on the provision of a uniform communication mechanism and the published interfaces for access to shared resources.
- Can be constructed by heterogeneous hardware and software (from different vendors, if possible). But for a system to work properly, the conformance of each component to the published standard must be tested and verified carefully.

Therefore, openness cannot be achieved if the specification and documentation of the key software of the interfaces of a system's components are not made available to software developers.

3. **Security:** Security of information resources available and maintained in a distributed system is very important. Security for information resources has following components:

- **Confidentiality:** Protection against disclosure to unauthorized individuals.
- **Integrity:** Protection against corruption or alteration.
- **Availability:** Protection against interference with the means to access the resources.

A firewall can be used as a barrier in an intranet to protect it from outside users, but it does not ensure the appropriate use of resources by users within the intranet.

To send sensitive information in a message over a network in a secure manner is a challenge. But security is not just a matter of hiding the contents of messages, it also involves knowing the identity of the user. Therefore, the second challenge is to identify the remote user. Both of the challenges can be met by using encryption techniques. Encryption is used to provide protection to shared resources and to keep sensitive information secure when it is transmitted in messages over a network. However, the two security challenges that have not yet been fully met are as follows:

- **Denial-of-Service Attack:** One more security problem is that a user may wish to disrupt a service for some reasons. This can be achieved by bombarding the service with a very large number of pointless requests that are not used by other users. This is called a Denial-of-Service (DoS) attack.
- **Security of Mobile Code:** Mobile code needs to be handled with care. For someone who receives an executable program as an e-mail attachment, the possible effects of running the programs are unpredictable.

4. **Scalability:** In a distributed system, particularly in a wide area distributed system, scalability is important. According to Neuman, a system is said to be scalable if: "It can handle the addition of users and resources without suffering a noticeable loss of performance or increase in administrative complexity."

A system is said to be scalable if it will remain effective when there is a significant increase in the number of users and resources. In designing a scalable distributed system, there are following challenges:

- Controlling the cost of physical resources
- Controlling the performance loss
- Preventing software resources running out
- Avoiding performance bottlenecks

5. **Failure Handling:** When there is any fault in hardware or software, programs may produce incorrect results or they may halt before they have completed the intended computation.

In a distributed system, failures are partial, that is, some components may fail while others continue functioning. Therefore, failure handling is difficult. Following are the techniques that deal with failures:

- **Detecting Failure:** Some of the failures can be detected. For example, to detect corrupted data in a message or a file, checksum can be used. A checksum, also known as hash sum, is a small size data computed from a random block of digital data which is used to detect errors while transmitting data.

- **Masking Failures:** Some of the failures that have been detected can be made less severe or hidden. For example, a file data can be written on a pair of disks so that if one gets corrupted the other one is still correct.
- **Tolerating Failure:** To detect and hide all the failures in a large network is not possible. The clients of such a network can be designed to tolerate failures, which generally involve the users tolerating them as well.
- **Recovery from Failure:** The design of software is involved in recovery, so that the state of permanent data can be "rolled back" or recovered after a server has crashed.
- **Redundancy:** By using redundant components, services can be made to tolerate failures. For example, in the DNS (Domain Name Server), every name table is replicated to at least two different servers.

6. **Concurrency:** In a distributed system, both the applications and the services provide resources that can be shared by clients. Therefore, there is a possibility that several clients will attempt to access the shared resource at the same time. The process that is used to manage a shared resource could take one client request at a time. But this approach limits the throughput. Therefore services and the applications usually allow multiple client requests to be processed concurrently. To keep an object safe in a concurrent environment, its operations must be synchronized in such a manner that its data remains consistent. This can be achieved by a technique such as *semaphores*.

7. **Transparency:** The aim of transparency is to make certain aspects of distribution invisible to the application programmers.

   - **Access Transparency:** This enables the local and remote resources to be accessed using identical operations.
   - **Location Transparency:** This enables the resources to be accessed without knowledge of their location.
   - **Concurrency Transparency:** This enables several processes to operate concurrently using shared resources.
   - **Replication Transparency:** This enables multiple instances of resources to be used to increase reliability and performance without having knowledge of the replicas by users or application programmers.
   - **Failure Transparency:** It allows the users and the application programs to complete their tasks, even though they have the failure of hardware or software components.
   - **Mobility Transparency:** The mobility of the resources and clients within a system will not affect the operation of users or programs.
   - **Performance Transparency:** It allows the system to be reconfigured to improve performance as loads vary.
   - **Scaling Transparency:** It allows the system and applications to expand in scale, and will not change the system structure or the application algorithms.

⊙ Self-Quiz

Q1. What are the goals of distributed system? List them.
Q2. What is transparency? Explain its types.
Q3. What is scalability? Why it is important for designing a distributed system?

EXAMPLES OF A DISTRIBUTED OPERATING SYSTEM

The three most widely used and familiar examples of distributed systems are as follows:

1. The Internet
2. Intranets
3. Mobile and ubiquitous computing

Internet

The Internet is a computer network made up of thousands of networks worldwide as shown in Figure 13.1. Interaction among programs running on the computers (those who are connected to the Internet) is held by passing messages. Therefore the major technical achievement is the Internet communication mechanism.

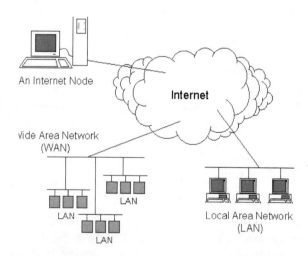

FIGURE 13.1. Collection of computer networks: Internet

The Internet is also considered as a distributed system. Through the Internet users are enabled to use services of World Wide Web (WWW), e-mail, file transfer, and so on. Internet Service Providers (ISPs) are the companies which provide modem links and some other types of connection to individual users and small organizations, which enable them to access services anywhere in the Internet. It also provides local services, such as

Web hosting and e-mail. For linking the intranets together, a backbone is used. A backbone network is a network link with high transmission capacity which employs satellite connections, fiber optic cables, and other high bandwidth circuits.

Multimedia services are also available with the Internet, which enables the users to access audio and video data including video conferences and to hold phone, music, radio, and television channels.

Intranet

The portion of the Internet that is separately administered is known as an intranet, which has a boundary that can be configured to enforce local security policies. Pictorial representation of an intranet is shown in Figure 13.2. Figure 13.2 illustrates a typical corporate intranet in which A represents firewalls, B represents LANs resources that connect the centrally controlled and managed Local Area Networks (LANs) to the intranet (bridges and routers), C represents the setting done with dial-up Internet access feature for various workstations, and D represents mainframe access control. The term dial-up Internet access represents a form of Internet access that uses the facilities of the Public Switched Telephone Network (PSTN) to establish a dialed connection to an Internet Service Provider (ISP) via telephone lines, where it is composed up of several LANs (Local Area Networks) connected by a backbone connection.

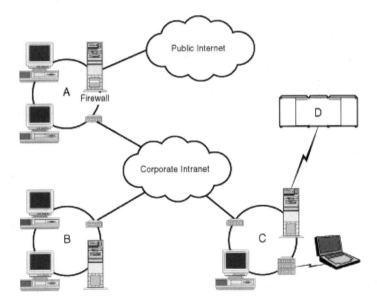

FIGURE 13.2. An intranet setup

The router is used to connect the intranet with the Internet, in which the users inside the intranet are allowed to use the services elsewhere, such as World Wide Web (WWW) or e-mail.

A firewall is used to protect an intranet from unauthorized use by malicious users. A firewall can be implemented by incoming messages and outgoing messages. The main issues that arise in the designing of components that are used in intranets are:

- File services: Needed to enable the user to share data.
- Firewalls with security mechanism are required when resources are shared among internal and external users.
- Software installation cost and support is an important issue. By using system architectures, such as network computers and their client, cost can be reduced.

Mobile and Ubiquitous Computing

Due to advancement in technology, small and portable computing devices are integrated with distributed systems (See Figure 13.3). These devices are:

- Laptop computers
- Handheld devices, such as pagers, mobile phones, Personal Digital Assistants (PDA), video and digital cameras, and so on.
- Wearable devices, such as smart watches.
- Devices that are embedded in appliances, such as wi-fi (Wireless Fidelity) systems, refrigerators, washing machines, and cars.

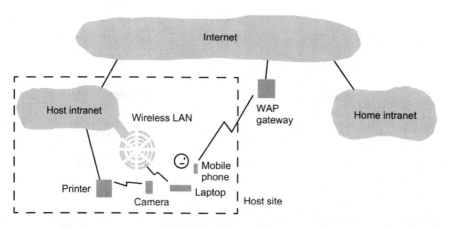

FIGURE 13.3. Portable and handheld devices in a distributed system

Figure 13.3 represents the portable and handheld devices in a distributed system in which the function of the WAP (Wireless Application Protocol) gateway is to sit between mobile devices using the WAP protocol and the World Wide Web (WWW), passing pages from one to the other that works much like a proxy. Wireless application protocol is widely used for mobile Internet service which can be used on a handheld device, such as a mobile phone or PDA.

Mobile computing becomes possible because many of these portable devices have the ability to connect with the networks in different places very conveniently. In mobile computing (also known as *nomadic computing* [Kllinrock 1997]) users can access the resources (via portable devices) while they are on the move or at a visitor's location.

Ubiquitous computing (Weiser 1993), also known as pervasive computing, is the harnessing of many small cheap computational devices that are present in the user's physical environment, such as in the home, office, or elsewhere.

Since mobile users can benefit from computers that are everywhere, there is overlap among ubiquitous and mobile computing. But in general, they both are very different. In ubiquitous computing users could get an advantage if they remain in a single environment, whereas in mobile computing users can profit if only conventional, discrete computers and devices are used.

Use of portable and handheld devices in a distributed system is shown in Figure 13.3, where a user is visiting to a host organization. Here three forms of wireless connections can be accessed by the user, i.e., (1) A laptop is connected to the host's wireless LAN, (2) A mobile phone is connected to the Internet using WAP (Wireless Application Protocol) through a gateway, and/or (3) A digital camera is connected to a printer over an infra-red link.

⊙ Self-Quiz

Q1. **What is the Internet? Illustrate.**
Q2. **What is intranet? Write the main issues that arise in the designing of components used in the intranet.**
Q3. **Give examples of a distributed system. Explain them.**

ADVANTAGES AND DISADVANTAGES OF A DISTRIBUTED SYSTEM

Some significant advantages of distributed systems are as follows:

■ **Resource Sharing:** Whenever a computer requests a service from another computer, it simply transmits an appropriate request to that computer over a communication network. Such communication takes place by sharing resources among computers over the network.

■ **Enhanced Performance:** Providing rapid response time and higher system throughput is the ability of distributed system and this is mainly because of the fact that many tasks are executed concurrently at different computers. Moreover, to improve the response time, a load distributing technique is used by the distributed system. In a load distributing technique, tasks are transferred from heavily loaded computers to lightly loaded computers.

■ **Improved Reliability and Availability:** Improved reliability and availability can be provided by a distributed system because failure of few components of the system does not affect the availability of the rest of the system. The distributed system can also be made fault tolerant, through the replication of data and services (processes that provide functionality).

■ **Modular Expandability:** New resources, either hardware or software, can be easily be added in a distributed system environment, without replacing it with the existing resources.

Although the distributed system has many advantages, it also has some disadvantages, which are as follows:

■ **Software:** This is the worst problem for a distributed system. At present, very little software is available for a distributed system, because designing, implementing, and using such distributed software is a challenging task to date.

■ **Network:** A second problem arises in a distributed system due to its communication network. When two or more devices are communicating, it is possible that they lose messages over a communication network, which could be recovered by special software. Using this software could overload the system and as a result, the network could be saturated. This situation can be handled either by replacing the current network or by adding one more network. Both the methods involve great expense.

■ **Security:** Security is always a major concern for any environment. Sharing data easily is an advantage, but it is also a disadvantage for a distributed system. People can very easily access the data (either for their use or not) all over the network, which affects the security features.

KEY TERMS

■ **CORBA:** Common Object Request Broker Architecture; provides general interface standards that can be supported by different programming languages.

■ **Ethernet:** A family of computer networking technologies for local area networks.

■ **FDDI:** Fiber Distributed Data Interface; a standard for data transmission in a local area network.

■ **Intranets:** A computer network that uses Internet Protocol technology to share information, operational systems, or computing services within an organization.

■ **Memory:** A temporary storage area.

■ **NUMA:** Non-uniform Memory Access; a memory design used in multi-processing, where the memory access time depends on the memory location relative to the processor.

SUMMARY

Distributed systems are everywhere. The Internet enables the users to access its services throughout the world. Each organization manages its intranet. Small distributed systems can be built from mobile computers and other small devices that can be attached to a wireless network. For constructing a distributed system, resource sharing is the main motivating factor. Many challenges are there for constructing a distributed system.

POINTS TO REMEMBER

- A distributed system is composed of several computers that do not share a common memory or a clock.
- The only way to communicate in a distributed system is by sending messages through the network.
- The requirement of a distributed system came from the fact of sharing resources among users.
- *Service* is the term used for a distinct part of a computer system that manages a collection of related resources and presents their functionality to the users and application through a well-defined set of operations.
- *Server* refers to a running program on a networked computer which accepts request messages from programs that are running on the other computers (i.e., client) to perform a service and to provide appropriate responses by replying to messages. A complete interaction between client and server is called a *remote invocation*.
- *Middleware* is applied to the software layer, which provides a programming abstraction as well as masking the heterogeneity of the underlying network, hardware, operating systems, and programming languages.
- *Mobile code* is code that can be sent from one computer to the other computer and run at the destination.

REVIEW QUESTIONS

Q1. What are the advantages of distributed systems?
Q2. What are the characteristics of distributed systems?
Q3. Differentiate between the Internet and the intranet.
Q4. Explain various examples of distributed systems.
Q5. What are the challenges involved in distributed systems? Explain them.
Q6. Discuss the issues that are relevant to the understanding of distributed, fault tolerance systems.
Q7. Write the functions of semaphores.
Q8. Describe the types of transparency and their features.

CHAPTER 14

LINUX

Purpose of the Chapter

This chapter explains the different components of the Linux operating system. It also explains how these components are connected with each other and how they work together to provide a convenient environment to the user. It also explains the architecture of the Linux operating system and its various communities.

INTRODUCTION

Linux is an operating system which provides a convenient platform to the computer users so that they can execute their applications and perform required functions in an efficient manner. The Operating System (OS) passes instructions from an application to the computer's central processing unit (CPU). The processor performs its processing on the instructions which come from the user's applications, its output is generated, and the desired result is sent back to the application through the operating system. Linux is also considered as one of the popular versions of the UNIX operating system.

WHAT IS LINUX?

Linux is considered an open source operating system as its source code is freely available. This operating system is free to use. Linux was designed considering UNIX compatibility. Its functions list is very similar to that of the UNIX operating system. As an open operating system, Linux is developed in collaboration. This means not a single company is solely responsible for its development or ongoing support. Many companies participate in its research and development. As many companies and individuals are involved in the creation, research, and development of Linux, it helps in shrugging the burden and also results in an effective ecosystem. In the development of a single Linux kernel, a team comprising more than 1,000 developers, from at least 100 different companies, works together. The logo of the Linux operating system is shown in Figure 14.1.

FIGURE 14.1. Logo of Linux operating system

HISTORY OF LINUX

On August 25, 1991, an announcement was made by a computer science student named Linus Torvalds to the Usenet group. In the words of Linus Torvalds, *"I'm doing a (free) operating system (just a hobby, won't be big and professional like gnu) for 386(486) AT clones. This has been brewing since April, and is starting to get ready. I'd like any feedback on things people like/dislike in minix, as my OS resembles it somewhat same physical layout of the file-system (due to practical reasons) (among other things)."*

In his announcement Torvalds used a name of an operating system which is "Minix." He considered it as an alternative to the UNIX operating system. He used UNIX as a baseline or ground for the new operating system he wanted to create. He wanted his creation to be open source. In his announcement the word "gnu" referred to the set of GNU tools which were first put together by Richard Stallman in 1983.

In starting, Torvalds designed and developed the core of the Linux operating system, which is the kernel of the Linux operating system. Kernel is just a piece of any operating system. It is not the whole of it. On the other side, Stallman's GNU tools were used to create an operating system but in that a piece of software, that is, kernel, is missing to make it a complete operating system. Hence Torvalds' Linux kernel along with GNU tools gave the starting point of an operating system which is now called Linux operating system. In 1991 Torvald in his announcement made the request publicly to let developers be involved as open source for better research and development of this operating system.

There were many charming features in Linux like portability to various platforms, its similarity with UNIX, and its free and open software license. These features attracted the UNIX developers and Linux gained rapid popularity amongst them.

By the end of the century, many commercial developer companies like VA Linux, TurboLinux, Mandrakelinux, Red Hat, and SuSE GMbH began to distribute Linux. In year 2000 IBM (International Business Machines) took the decision to invest $2 billion for the development and sales of Linux. This step and investment of IBM proved a huge point regarding the growth of Linux.

In the present time, Linux is considered as a multibillion-dollar industry. Because of its remarkable features like security and flexibility, hundreds of companies and governments are taking advantage of the operating system. Low licensing and support costs are major attractions in the use of Linux.

COMPONENTS OF THE LINUX OPERATING SYSTEM

The Linux operating system consists of three main components, which are as follows:

1. Kernel
2. System library
3. System utility

- **Kernel:** The core part of Linux is the kernel. All the major activities of this operating system are handled by the kernel. It is comprised of various modules. The kernel directly interacts with the hardware which lies below it. One of the main functions of the kernel is to provide abstraction. Provision of abstraction is required to hide low-level hardware complexities from the user and its application programs.
- **System library:** Special functions or programs with the help of application programs or system utilities which are able to access various features of the kernel is known as the system library. The modules present in the system library help in the implementation of functionalities of the operating system.
- **System Utility:** The third major component of the Linux operating system is the system utility programs. They are responsible for specialized, individual-level tasks.

The three major components of the Linux operating system and their placing are shown in Figure 14.2.

ARCHITECTURE OF THE LINUX OPERATING SYSTEM

Linux system architecture comprises mainly four layers which are as follows:

- **Hardware layer:** The first and bottom-most layer is hardware, comprising of all peripheral devices like Random Access Memory (RAM), hard disk drive (HDD), CPU, and input/output devices.

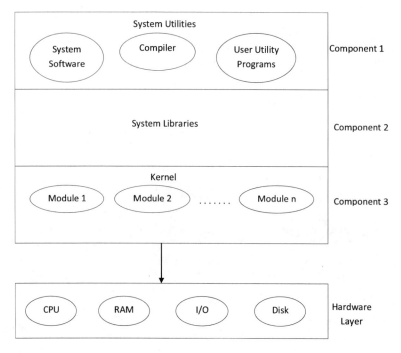

FIGURE 14.2. Components of the Linux operating system

- **Kernel:** The kernel is the core component of the operating system. It interacts directly with the hardware. It provides low-level services to upper-layer components.
- **Shell:** The shell is an interface to the kernel. It hides the complexity of the kernel's functions from users. Commands from user instructions are forwarded to the shell. In response, it executes the functions of the kernel.
- **Utilities:** There are various functions which are handled by an operating system. The programs performing all those functions are present in utilities.

The architecture of the Linux operating system is shown in Figure 14.3.

The two modes in Linux are kernel mode and user mode. Let us discuss each of them in brief:

- Kernel mode is considered as a special privileged mode in which kernel component code executes. In this mode, applications have full access to all resources of the computer. This code represents a single process. It executes in a single address space and there is no need of any context switching. That is why kernel mode is considered to be very efficient and fast. Kernel helps in the execution of each process. It is responsible for providing various system services to different processes. It provides protected access of hardware so that processes use the resources in a protected manner.

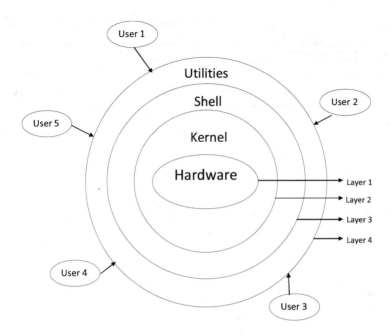

FIGURE 14.3. Architecture of the Linux operating system

■ User programs and other system programs work in user mode. They have no direct access to the system hardware. Also such programs cannot access kernel code. Such user programs and utilities have to use system libraries in order to access kernel functions to get the system's low-level tasks.

LINUX KERNEL SERVICES

The Linux kernel provides the following services:

■ The kernel supports all hardware drivers. A few drivers are compiled into the kernel, but most are loaded on demand. The kernel configuration determines the drivers compiled into the kernel, and drivers available for dynamic loading.

■ The kernel manages memory allocation to all processes and buffers. During the boot process, the kernel allocates much of the available memory to buffers. As processes demand memory, the kernel releases unused buffer space. The kernel always keeps a reserve area in memory, and, if necessary, will terminate processes that have too much memory. Killing a memory of process is the last resort, and occurs only when swapping no longer solves the problem with available memory.

■ The kernel manages all processes via a scheduler. The scheduler itself is configurable. Thus, a scheduler configuration for a desktop favors streaming of data to improve the performance of music and videos. The scheduler configuration for servers optimizes performance for server processes.

- Using the Virtual File System (VFS), a generic interface for all file systems, the file system drivers implement the code necessary to interface with each file system. Like device drivers, most file system drivers are loaded on demand.
- The kernel implements all the drivers necessary to support networking.
- File system security is part of the kernel, using firewalls and virus-checking software.

MAIN SYSTEM CALLS OF LINUX

The main system calls of the Linux operating system are as follows.

File Related System Calls

- Close - Close a file descriptor.
- Link - A new name is made for a file.
- Open - Open and possibly create a file or device.
- Read - Read from a file descriptor.
- Write - Write from a file descriptor.

Process-Related System Calls

- Execve - Execute program.
- exit - Terminate the calling process.
- getpid - Get process identification.
- setuid - Set user identity of the current process.
- Prtrace - Provides a way by which a parent process controls the execution of another process.

Scheduling Related System Calls

- sched_getparam - This call sets the scheduling parameters associated with the scheduling policy for the process.
- sched_get_priority_max - It helps in return of the maximum priority value. The priority value is to be used in scheduling algorithms.
- sched_setscheduler - Sets both the scheduling policy (e.g., FIFO or First In First Out) and the associated parameters for the process pid.

sched_rr_get_interval

Writes into the timespec structure pointed to by the parameter to? the Round Robin time quantum for the process.

Interprocess Communication (IPC) Related System Calls

- msgrcv -A message buffer structure is allocated to receive a message. The system call then reads a message from the message queue specified by msqid into the newly created message buffer.
- semctl -Performs the control operation specified by cmd on the semaphore set semid.

- Semop -Performs operations on selected members of the semaphore set semid.
- Shmat -This call attaches the shared memory segment to the data segment of the calling process.
- shmctl - It allows the user to receive information on a shared memory segment. It also sets the owner, group, and permissions of a shared memory segment, or deletion of a segment.

Socket (Networking) Related System Calls

- bind -Assigns the local Internet Protocol (IP) address and port for a socket. Returns 0 for success and – 1 for error.
- connect - Establishes a connection between the given socket and the remote socket associated with sockaddr.
- Gethostname -Returns local host name.
- send -Sends the bytes contained in the buffer pointed to by °msg over the given socket.
- Setsockopt -Sets the options on a socket.

Miscellaneous System Calls

- create_module - This system call helps in creation of a loadable module entry. It also reserves the kernel memory which is required to hold the module.
- fsync - It copies all in-core parts of a file to disk. It also waits until it gets a report from the device that all parts are on stable storage.
- query_module - Requests information related to loadable modules from the kernel.
- Time - Returns the time in seconds since January 1, 1970.
- vhangup - Simulates a hangup on the current terminal. This call arranges for other users to have a "clean" terminal at login time.

MAIN FEATURES OF THE LINUX OPERATING SYSTEM

The important features of the Linux operating system are explained as follows:

- **Portable:** An operating system is said to be portable if can work on different types of hardware in just the same way. The Linux kernel and its application programs support portability on all hardware platforms.
- **Open Source:** Linux source code is freely available because it is an open source operating system. It is a community-based development project. Many teams work in collaboration to increase the capability and functionality of the Linux operating system and it is continuously evolving.
- **Multi-User:** The Linux operating system is a multiuser system. This means multiple users can access the resources (like memory, RAM, application programs) of the system at the same time without any trouble.

- **Multiprogramming:** Linux is a multiprogramming system. This means multiple applications can execute at the same time and give results.
- **Hierarchical File System:** Linux provides a hierarchical file structure in which system files and user files are arranged in a hierarchical manner.
- **Shell:** Linux has an interpreter program called shell. It is used to execute various commands of the operating system. It can also be used to handle various types of operations.
- **Security:** Linux provides user security by using various authentication features. The different features are password protection, controlled access to specific files, encryption of data, and so on.

AREA OF USE OF LINUX OPERATING SYSTEM

Every operating system is used for different environments; for example, the Windows operating system is mainly used in personal computing devices like desktop and laptop computers. Operating systems, like Symbian, Android, and so on, are mainly for mobile devices such as phones. Mainframes and supercomputers, which are found in major scientific areas and academic and corporate labs, use specific operating systems such as AS/400 and the Cray OS.

The Linux operating system began as a server operating system. Now it has become useful as a desktop OS. It can also be used on all of devices from wrist watches to supercomputers.

DIFFERENCES BETWEEN LINUX AND UNIX OPERATING SYSTEMS

1. **The difference in approach:** There are two types of kernels, that is, monolithic and micro-kernel. Some groups of kernels fall in the monolithic category. Monolithic kernels are those which operate in one and only one process. They don't have any other process for any kind of task. Another category is called micro-kernel, in which the kernel is assigned one process, whereas other processes are present for performing side tasks like drivers, and so on. Linux supports the monolithic category with few exceptions in micro-kernel, whereas UNIX comes in the micro-kernel category.

2. **Loadable modules of the kernel:** In UNIX systems there are static links of new modules which are to be added or recently added. Linux is different in this aspect. Linux supports new additions of the modules on the fly. For example, drivers can be loaded dynamically whenever they are required. This feature of Linux is called loadable kernel modules (LKM). This feature helps the user to add any component dynamically without any need of re-compiling the kernel. Similarly, the unloading of modules can be performed. This adds flexibility to Linux.

3. **Kernel threads:** The kernel thread is an independent execution flow. It can be used to execute some user process or any kernel code. Threads always operate in the same address space, so it is not expensive to perform context switching on kernel threads in comparison to processes. This explains why UNIX-like

systems have kernel threads, whereas in Linux, kernel threads are used to execute kernel code.

4. **Multi-threading:** Multi-threaded applications are those in which there are multiple execution flows in a single application. These multiple flows are also known as threads. Threads are also called light weight processes. Nowadays all operating systems support multi-threading. UNIX supports LWP (Light Weight Process), which is kernel thread-based, whereas Linux handles thread in a different manner. In Linux, LWP are created by calling the clone () function. This function results in the creation of separate processes. The same work can also be performed by the fork () function, but clone () makes recently created processes share their memory, address space, and so on. Their working in a shared environment gave them the name "threads." In short, multi-threading is supported by both UNIX and Linux, but they differ in internal handling of it.

5. **Preemption and non-preemption:** The kernels which can preempt the currently executing process are called preemptive kernels. Mainly processes are executed on a priority number basis. If the currently executing process has a low priority and a high-priority process arrives, then it can interrupt the current process and start executing itself. Non-preemptive kernels are those which cannot forcibly interrupt the current process. In such cases a high-priority process has to wait. Linux-based operating systems support non-preemptive kernels while UNIX systems are fully preemptive. Linux-based real-time operating systems are found to be fully preemptive.

6. **Cost:** Linux can be freely distributed, downloaded freely, distributed through magazines, books, and so on. There are priced versions for Linux also, but they are normally cheaper than Windows, whereas UNIX has different cost structures according to vendors.

7. **User group:** The user of Linux is everyone, from home users to developers, whereas UNIX operating systems were developed mainly for mainframes, servers, and workstations.

8. **Usage**: Linux can be installed on a wide variety of computer hardware, ranging from mobile phones and tablets, whereas the UNIX operating system is used in Internet servers, workstations, and PCs.

9. **Manufacturer:** The Linux kernel is developed by the community, whereas three biggest distributions into UNIX are Solaris (Oracle), Advanced Interactive eXecutive (AIX) (IBM), and HP-UX Hewlett Packard UNIX.

THE FUTURE OF LINUX

In the present scenario, Linux is already considered as a successful operating system which is used in many different kinds of devices, but there are also many other technological areas where Linux is expanding. It is being installed on the system Basic Input/Output System (BIOS) of laptop and notebook computers. Such an environment can have Internet connectivity tools with browsing capabilities, and can help its users to work on the Internet. It removes the need to boot into their system's main operating system—even if that operating system is Windows.

At the same time, Linux can also can be used as Mobile Internet Devices (MIDs), for example, embedded devices like smart phones, PDAs, net book devices, small laptop-type machines, and so on.

Linux has already proved its competence in upcoming technology, that is, cloud computing. Linux enables cloud services such as Amazon's A3 to work with superior capability to deliver online applications and cater the needs of the users.

Linux is also successful when loaded in supercomputers, both in the High-Performance computing (HPC) and High-Availability (HA) areas. It is also competent in academic research in physics and bioengineering, and organizations in the financial and energy industries, which need reliable computing power to reach their goals.

The most popular Web 2.0 services on the Internet, like Twitter, LinkedIn, YouTube, and Google, rely on Linux as their operating system. As new Web services arrive in the future, Linux will increasingly be the platform that drives these new technologies.

RED HAT LINUX

In 1994, Bob Young and Marc Ewing founded the company "Red Hat." Its headquarters were formed in the United States, in North Carolina. From the very beginning the open source operating system, Linux, played a high role for the enterprise concept. The field of application reached from miniature devices over workstations up to server systems on Intel x86, Dec alpha, and Sun SPARC (Scalable Processor ARChitecture) systems. The strength of Red Hat Linux is its application in Internet and intranet. Extensive support, training, and training offers and huge support from IT companies helped in the growth of Red Hat. Red Hat touched about 15% by the gross income, with Linux distribution. Red Hat has a set of partnerships which are as follows:

1998:   Partnerships with Intel and Netscape
1999:   Partnerships with SAP (stands for Systeme, Andwendungen, Produkte in der Datenverarbeitung which has been translated to English, i.e., means Systems, Applications, Products in Data Processing), Oracle, IBM, Compaq, Dell and Novell

One of the very famous and customer approved versions of the Linux distribution is the Fedora project of Red Hat. This open source project is sponsored by Red Hat, but it is independent of the Linux Community. In May 2004 the Fedora core 2 for the x86-64 and i386 architecture was published, and used the Linux Kernel 2.6.

VERSIONS OF RED HAT LINUX

There are various versions of Linux. Table 14.1 summarizes the versions of Red Hat Linux.

TABLE 14.1. Versions of Red Hat Linux

Release Date	Kernel Version
1995	Red Hat Linux 1.0
1995	Red Hat Linux 2.0
1996 May	Red Hat Linux 3.0.3 (picasso), kernel 1.2
1996 October	Red Hat Linux 4.0 (colgate), kernel ?
1997 April	Red Hat Linux 4.2 (biltmore), kernel 2.0.30
1997 November	Red Hat Linux 5.0 (hurricane), kernel 2.0.32
1998 May	Red Hat Linux 5.1 (manhattan), kernel 2.0.34
1998 October	Red Hat Linux 5.2 (apollo), kernel 2.0.36
1999 April	Red Hat Linux 6.0 (hedwig), kernel 2.2.5
1999 September	Red Hat Linux 6.1 (cartman), kernel 2.2.12
2000 March	Red Hat Linux 6.2 (zoot), kernel 2.2.14
2000 August	Red Hat Linux 7.0 (guiness), kernel 2.2.16
2001 April	Red Hat Linux 7.1 (seawolf), kernel 2.4.2
2001 October	Red Hat Linux 7.2 (enigma), kernel 2.4.7
2002 May	Red Hat Linux 7.3 (vallhalla), kernel 2.4.18, ext3
2002 September	Red Hat Linux 8.0 (psyche), gcc 3.2, kernel 2.4.18
2003 April	Red Hat Linux 9.0 (shrike), gcc 3.2.1, kernel 2.4.20
2003 November	Fedora Linux Core 1
2004 May	Fedora Linux Core 2
2004 November	Fedora Linux Core 3
2005 June	Fedora Linux Core 4, kernel 2.6.9
2006 March	Fedora Linux Core 5
2006 October	Fedora Linux Core 6
2007 May	Fedora Linux Core 7
2007 November	Fedora Linux Core 8
2008 May	Fedora Linux Core 9
2005 January	CentOS 3.4
2006 March	CentOS 4.0
2006 March	CentOS 4.3
2007 December	CentOS 4.6
2008 September	CentOS 4.7, based on open source code of RHEL 4.7
2007 Dec.	CentOS 5.1
2008 June	CentOS 5.2, based on open source code of RHEL 5.2

MISCELLANEOUS POINTS REGARDING LINUX

Linux Environment

The desktop environment consisting of windows, menus, and dialog boxes are actually separate layers of an operating system. These layers provide the user-oriented Graphical User Interface (GUI) that helps the users to work conveniently with applications in the operating system. In Linux, there are many choices regarding the desktop environment, something that Linux allows users to decide. This feature is not present in the Windows operating system.

The Linux operating system and its kernel consists of tools and code libraries which helps application developers to work with these environments, for example, GIMP Tool Kit GTK+ for GNOME GNU Image Manipulation Program Tool Kit, Qt for KDE, or K Desktop Environment.

Linux Distribution

This is the highest layer of the Linux operating system. It is considered as a container for all the aforementioned layers. Distributor decides which kernel, operating system tools, environments, and applications are to be included for the use of users.

Distributions of Linux can be maintained by individuals and by commercial organizations. A distribution can be installed using a Compact Disk (CD) that contains distribution-specific software for initial system installation and configuration. For the users, most popular distributions offer mature application management systems that allow users to search, find, and install new applications with just a few clicks of the mouse. At present there are over 350 distinct distributions of Linux.

Linux Community

Developer and user communities are the two Linux communities. One of the most charming features of Linux is its accessibility to developers. Linux is an open source operating system, hence anyone with the required skills can improve the code of Linux. This results in and helps in its development and improvement. Developer communities volunteer to maintain and support whole distributions, such as the Debian or Gentoo Projects. Novell and Red Hat also support community-driven versions of their products, like openSUSE and Fedora.

There is an increase in the number of developers and companies who want to be the part of development of Linux. Hardware vendors want to ensure that Linux supports their products well, making those products attractive to Linux users. Embedded systems vendors, who use Linux as a component in an integrated product, want Linux to be as capable and well-suited to the task at hand as possible. Distributors and other software vendors who base their products on Linux have a clear interest in the capabilities, performance, and reliability of the Linux kernel. Other developer communities focus on different applications and environments that run on Linux, such as Firefox, OpenOffice. org, GNOME, and KDE.

Linux Development

The Linux operating system comprises several development languages. A very large percentage of the distributions' code is written in either the C (52.86%) or C++ (25.56%) languages. The rest of the Linux code uses Java, Perl, and LISP (LISt Processing), rounding out the rest of the top five languages.

Most of Linux kernel code is written in C, but other languages are also used making it more heterogeneous than other operating systems.

The kernel community has evolved its own distinct ways of operating which allow it to function smoothly, resulting in a good quality product in an environment where thousands of lines of code are being changed every day. The development community is very helpful to those who are trying to learn and develop Linux kernel. While many Linux developers still use text-based tools, such as Emacs or Vim to develop their code, Eclipse, Anjuta, and NetBeans all provide more robust integrated development environments for Linux.

SUMMARY

In this chapter, you have learned the basics of the Linux operating system. Linux is a UNIX-like and mostly Portable Operating System Interface for UniX or POSIX-compliant operating system which was originally developed as a free operating system for Intel x86-based PCs. One of the prime features of Linux is that it also runs on embedded systems, which includes mobile phones, tablets, network routers, televisions, and video game consoles. You have also learned about some popular Linux distributions, which include Debian, Ubuntu, Linux Mint, Fedora, openSUSE, Arch Linux, and the commercial Red Hat Enterprise Linux and SUSE Linux Enterprise Server.

POINTS TO REMEMBER

- Shell is an interface to the kernel, and it hides the complexity of the kernel's functions from users.
- Kernel mode is considered as a special privileged mode in which kernel component code executes.
- Linux source code is freely available because it is an open source operating system.
- Linux has an interpreter program called shell.
- Operating systems like Symbian, Android, and so on are mainly for mobile devices such as phones.
- Kernel thread is an independent execution flow which can be used to execute some user process or any kernel mode.
- In Linux, Light Weight Process (LWP) is created by calling the clone() function.

■ The Linux operating system and its kernel consists of tools and code libraries which help application developers to work with these environments, for example gtk+ for GNOME and Qt for KDE.

■ The most popular Web services on the Internet, like Twitter, LinkedIn, YouTube, and Google, rely on Linux as their operating system.

REVIEW QUESTIONS

Q1. What is Linux? Explain.

Q2. Describe the components of the Linux operating system.

Q3. Write the prime function of the kernel.

Q4. Discuss the architecture of the Linux operating system with the help of illustration.

Q5. Write a short note on Linux kernel services.

Q6. What is multi-threading?

Q7. Describe the Linux environment.

REFERENCES

[1] http://www.linuxfoundation.org/what-is-linux

[2] http://www.tellmeaboutlinux.com/content/linux-architecture

[3] http://dator8.info/pdf/LINUX/4.pdf

[5] http://www.operating-system.org/betriebssystem/_english/bs-redhat.htm

[6] http://www.redhat.com/en/about/blog/red-hat-enterprise-linux-the-original-cloud-operating-system

CHAPTER **15**

MOBILE OPERATING SYSTEMS

Purpose of the Chapter

Mobile phone technology has seen rapid changes in its use, look, design, and services. Most of the support to the technology is provided by an operating system which is specially designed for mobile phones. In the past ten to fifteen years, there have been extensive changes in the design and support provided by operating systems. The purpose of this chapter is to explain the main operating systems which are in extensive use at present. This chapter will explain the basic concepts of the Android, IOS, and Windows 8 operating systems.

INTRODUCTION

With the advent of mobile phone technology, the world's geographical boundaries are shrinking. With advancements in the technology and a competitive market, the cost of usage is diminishing. This is resulting in an increase of users of mobile phones. Nowadays these devices are not just for making phone calls, but are used as a multipurpose device. They have changed our way of life and have now become an essential tool in our day-to-day life. Phones are used for many purposes, like taking pictures, making videos, playing music, browsing, chatting, and so on. For the provision of various facilities and for the support for various applications, various types of operating systems are required.

ANDROID OPERATING SYSTEM

Various types of operating systems are in use to provide various services on mobile phones. One such operating system which is in extensive use is the Android operating system. It is an open source and Linux-based operating system. It is not only an operating system, but also comprises many applications. Android Inc. was founded in Palo Alto, California, United States, by Andy Rubin, Rich Miner, Nick Sears, and Chris White in 2003. Later Android Inc. was acquired by Google in 2005.

Various Versions of Android

Since its original release, there have been various updates in the original version of Android.

The various versions and support provided by each version of Android is given in Table 15.1, and Android's logo is shown in Figure 15.1.

TABLE 15.1. Various Versions of the Android Operating System

S.No.	Version	Release Year	Support provided
1	Android 1.1	Feburary 2009	Support for saving attachments in Multimedia Message Service (MMS)
2	Android 1.5 Cupcake	April 2009	Support for Bluetooth; Uploading pictures using application Picasa
3	Android 1.6 Donut	September 2009	Google support
4	Android 2.0/1 Eclair	October 2009	Support for Hyper Text Markup Language (HTML)
5	Android 2.2 Froyo	May 2010	Support for Adobe Flash
6	Android 2.3 Gingerbread	December 2010	Support for extra large screen size
7	Android 3.0 Honeycomb	May 2011	Support for video chat
8	Android 4.1 Jelly Bean	June 2012	Improved user interface

FIGURE 15.1. Logo of the Android operating system

Architecture of the Android Operating System

The Android is comprised of different layers. Each layer consists of a group of many program components. In all, it includes an operating middleware and important applications. In Android architecture, each layer provides services and support to the layer which is present above it. Android architecture is shown in Figure 15.2.

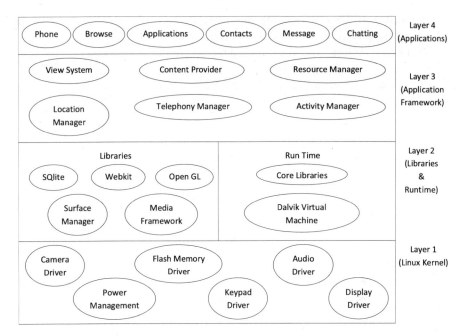

FIGURE 15.2. Various layers of Android architecture

Layer 1 – Linux Kernel

The very first layer in the architecture is the Linux kernel. The Android OS is built on top of the Linux 2.6 Kernel. The Linux kernel interacts with the hardware. It is comprised of all the required hardware drivers. *The programs which control and communicate with the hardware are called drivers.*

For example, to use the Bluetooth service, the device should have Bluetooth hardware in it. Hence the kernel must have a Bluetooth driver which can communicate with the Bluetooth hardware. The Linux kernel also acts as an abstraction layer between the hardware and other software layers. This kernel consists of various essential drivers for keypad, display, power management, audio, video, camera, and so on.

Layer 2 – Libraries and Runtime

The second layer in the architecture is of libraries and runtime. This layer helps the device to handle different types of data. C or C++ language is used to write the code for various libraries. These are specific to a particular hardware.

Some of the important libraries are:

■ **Surface manager:** It is used for compositing window manager with off-screen buffering. Off-screen buffering means in place of drawing directly into the screen, the drawings go to the off-screen buffer. There it is combined with other drawings and form the final screen the user will see. This off-screen buffer helps in providing the transparency of windows.

- **Media framework:** Media framework provides different media codes which help in recording and playback of different media formats.
- **SQLite:** SQLite is the database engine which is used for data storage purposes.
- **WebKit:** It is the browser engine used to display HTML content.
- **OpenGL:** It is used to provide two-dimensional (2D) or three-dimensional (3D) graphics content to the screen.

Android Runtime

Android runtime consists of the Dalvik Virtual machine and Core Java libraries.

- **Dalvik Virtual Machine:** It is a type of Java Virtual Machine (JVM) which is used to run applications in Android-based devices. It is optimized for low processing power and low memory environments. The Dalvik VM allows multiple instances of virtual machine to be created simultaneously. It provides security, isolation, memory management, and threading support.
- **Core Java Libraries:** These libraries provide the functionalities of Java Standard Edition (SE) libraries.

Layer 3 – Application Framework

Layer 3 is of the application framework. It consists of blocks with which applications directly interact. These blocks are actually programs. They manage the basic functions of mobile phones like resource management, voice call management, and so on. Important blocks of Application framework are:

- **Activity Manager:** It manages the activity life cycle of applications.
- **Content Providers:** It helps in the data sharing between applications.
- **Telephony Manager:** It manages all voice calls.
- **Location Manager:** It helps in location management, by using Global Positioning System (GPS) or cell towers.
- **Resource Manager:** It manages various types of resources which are used in various applications.

Layer 4 – Applications

The top layer in the Android architecture is applications. In this layer, all the applications are present. Various standard applications come pre-installed with every mobile phone, for example:

- Short Message Service (SMS)
- Dialer
- Web browser
- Contact manager

Features of the Android Operating System

As the market is competitive, this operating system has various features to fulfill the user's needs. Various features and a description of each feature of the Android operating system are given in Table 15.2.

TABLE 15.2. Various Features and Their Description of the Android Operating System

S.No.	Feature	Description
1	User Interface	This OS provides a beautiful and easy-to-use user interface.
2	Connectivity	It provides connectivity to various technologies, for example, GSM, CDMA, UMTS, Bluetooth, Wi-Fi, LTE, NFC, and WiMAX.
3	Storage	For data storage purposes, "SQLite," a lightweight relational database, is used.
4	Media support	It provides full media support in various forms like H.263, H.264, MPEG-4 SP, AMR, AAC, MP3, MIDI, JPEG, PNG, GIF, and BMP.
5	Messaging	Full support for SMS and MMS.
6	Web browser	It is based on the open-source WebKit layout engine, coupled with Chrome's V8 JavaScript engine supporting HTML5.
7	Multitasking	Users can easily switch from one task to another. Also users can run various applications simultaneously.
8	Multi-language	It supports the text of various languages.
9	GCM	It supports Google Cloud Messaging (GCM), which is a service that helps developers to send short message data to their users on Android devices.

The full forms of the terms used in Table 15.2 are as follows:

GSM: Global system Mobile
CDMA: Code Division Multiple Access
UMTS: Universal Mobile Multiple Access
Wi-Fi: Wireless Fidelity
LTE: Long Term Evolution
NFC: Near Field Communication
WiMAX: Worldwide Interoperability for Microwave Access
MPEG-4: Moving Picture Experts Group-4
AMR: Adaptive Multi-Rate
AAC: Advanced Audio Coding
MIDI: Musical Instrument Digital Interface
JPEG: Joint Photographic Experts Group
PNG: Portable Network Graphics
GIF: Graphic Interchange Format
BMP: Bitmap Picture

IOS

The mobile operating system especially designed for Apple-manufactured devices is 0069s iOS. It runs on the iPhone, iPad, iPod Touch, and Apple TV. This operating system is derived from Mac OS X. IOS is a UNIX-like OS. It is underlying software which allows iPhone users to interact with their phones using touch gestures such as swiping, tapping, and so on.

Various Versions of iOS

From its original release there have been various updates in the original version of iOS.

The various versions and support provided by each version of iOS are given in Table 15.3, and a logo of iOS is shown in Figure 15.2.

FIGURE 15.2. Logo of iOS

TABLE 15.3. Various Versions of iOS

S.No.	Version	Release Year	Support provided
1	**iOS 5**	October 12, 2011	• Support for iCloud and iTunes Match • iMessage
2	**iOS 5.1**	March 7, 2012	• Support for third-generation iPad • Update AT&T (American Telephone & Telegraph Company) network indicator to show 4G in many locations
3	**iOS 5.1.1**	May 7, 2012	• Solved a problem with 2G/3G network switching
4	**iOS 6**	September 19, 2012	• Turn-by-turn directions • Support for FaceTime over cellular networks
5	**iOS 6.1.5**	November 14, 2015	• Bug fix for 4th generation iPod touch
6	**iOS 7**	September 18, 2015	• iTunes in the Car • Airdrop
7	**iOS 7.1**	March 10, 2014	• Improvements to the Touch ID fingerprint scanner on the iPhone5S • Enhancements to the Calendar app and accessibility features

Architecture of iOS

The iOS mainly comprises of four abstraction layers. These layers are Core OS Layer, Core Services Layer, Media Layer, and Cocoa Touch Layer. In iOS, each layer provides services and support to the layer which is present above it. Each layer contains fundamental services and technologies.

iOS is shown in Figure 15.3.

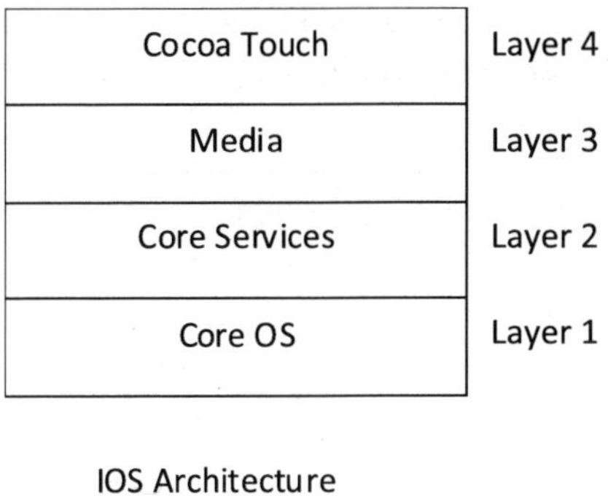

Cocoa Touch	Layer 4
Media	Layer 3
Core Services	Layer 2
Core OS	Layer 1

IOS Architecture

FIGURE 15.3. Various layers of iOS

Layer 1 – Core OS Layer

This layer provides all the features which are required to provide interaction with external hardware. Besides interaction with hardware, this layer provides much other support which is required for:

- Networking
- File system access
- Standard I/O (Input/Output)
- DNS (Domain Name System) services
- Locale information
- Memory allocation

Layer 2 – Core Services Layer

This layer provides services which are required by upper layers. The various support and services are as follows:

- Low-level networking
- Persistence in the SQLite database
- Date/time, threading
- GPS, cell, and Wi-Fi locations
- Core telephony framework
- Get cell provider information and receive call events
- In-App purchases

Layer 3 – Media Layer

This layer provides the necessary technologies for graphics, audio, and video.

- Core graphics
- 2D vector and raster graphics
- Core animation
- View animation
- Core image (Objective-C)
- Image and video manipulation, filters

- OpenGL ES and GLKit
- Hardware-accelerated 3D graphics
- Core text

Layer 4 – Cocoa Touch Layer

This layer consists of frameworks that are required in creating an application.

The various frameworks are as follows:

- **Address Book** – It helps in creating, saving, and managing contacts.
- **Event Kit UI** – It supports the creation of events and their reminders in calendar.
- **Game Kit** – It supports multiplayer games. It can add multiplayer capabilities to the games with peer-to-peer network using a Bluetooth connection. No pairing is required.
- **iAd** – This is mainly used for advertisements. It allows users to interact with advertising without having to leave their app.
- **Map Kit** – It adds map features directly into your application. It uses projected map coordinates instead of regions.
- **Message** – It supports mail and SMS.
- **UIKit** – The UIKit provides the basic tools required to implement graphical, event-driven apps in iOS.

WINDOWS 8

An operating system developed by Microsoft to be used in mobile phones is Windows 8. It has been considered as the next generation of Microsoft Windows operating systems. It is a part of the Windows NT family of operating systems designed by Microsoft. Its development started in 2009 even before the release of its predecessor, Windows 7. Many of its features are similar to Windows 7, but it has been specifically designed to be more mobile friendly.

Various Versions of the Windows Operating System

From its original release, there have been various updates in the original version of the Windows operating system. The various versions and support provided by each version of the Windows operating system are given in Table 15.4, and a logo of the Windows operating system is shown in Figure 15.4.

FIGURE 15.4. Logo of the Windows Operating System

TABLE 15.4. Various Versions of the Windows Operating System

S. No.	Version	Release Year	Support provided
1	Windows 3.1	April 1992	• Supports Graphical User Interface (GUI). • Based to run on the DOS (Disk Operating System). • 16-bit operating system that was designed for personal computers.
2	Windows NT	27 July 1993	• 32-bit operating system. • Multiprocessor and multiuser operating system. • Removed the dependencies on DOS. • Full rely on the NT Kernel.
3	Windows 95	24 August 1995	• Supports better user interface. • Supports network. • Plug-and-play feature (simply connect devices to the system. Removes the need of configuration).
4	Windows 2000	17 February 2000	• Supports integrated Web capabilities • Support for mobile computers and hardware devices. • It provides an easy way for business user to connect to the Internet anywhere and anytime.
5	Windows ME	14 September 2000	• Windows Millennium or ME. • Mainly designed for home users. • Supports user to connect users' computers at home to create a local area network with better Internet connectivity. • Supports multimedia content, such as photos, videos, and music.
6	Windows XP	25 October 2001	• Supports GDI+ graphics subsystem. • Better image management. • Start Menu and Taskbar improvements. • Integrated networking. • Multimedia support.
7	Windows 7	22 October 2009	• Faster sleep and resume. • Less memory needs. • Supports remote media streaming.
8	Windows 8	26 October 2012	• Multitasking with applications. • Supports free online storage.

Architecture of the Windows Operating System

The latest architecture of the Windows operating system is the Windows 2000 architecture. All newer versions of the Windows OS follow same architecture as Windows 2000. It consists of various layers as given below:

- Hardware Abstraction Layer (HAL)
- Kernel
- Executive services
- Environment subsystems

Each layer contains various services. The architecture of the Windows OS is shown in Figure 15.5.

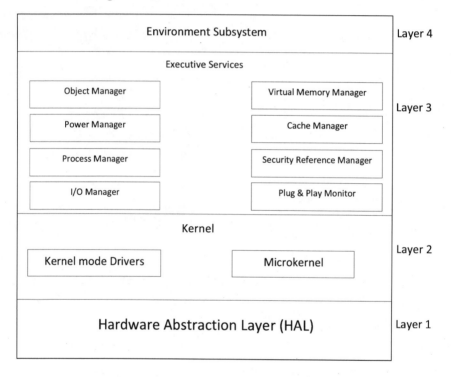

Windows Architecture

FIGURE 15.5. Various layers of Windows Operating System

Layer 1-Hardware Abstraction Layer

HAL is a thin layer of software. It abstracts the hardware differences from the operating system. This layer consists of a set of routines which supports a program to get direct access to the hardware resources. The Hardware Abstraction Layer (HAL) helps in providing transparency between the hardware and rest of the operating system. It allows Windows to be portable from one hardware platform to another.

Layer 2-Kernel

The kernel is placed to schedule the task which is to be performed by the CPU. When multiple processors are present in a computer, then the kernel synchronizes all the tasks among processors. By doing so it optimizes the processor performance. This layer consists of two modules:

■ Kernel-mode device drivers
■ Microkernel

Kernel-mode device drivers are used to interact with hardware components. These drivers have some system routines that represent all the system devices seen as a file object in the input/output manager for the user, and the I/O manager can view them as device objects. These drivers exist in three levels:

■ High level–The high level relies on the intermediate level, which contains function drivers.
■ Intermediate–The intermediate drivers rely on the lower level. This level also consists of the Windows Driver Model (WMD).
■ Low level drivers–The lowest level directly controls the hardware. This level is not dependent on intermediate or high-level drivers.

The Microkernel is a collection of programs. It helps in performing tasks like (a) Address space management, (b) Thread management, and (c) Inter-process Communication (IPC).

The Microkernel and windows kernel work together to make the operating system more efficient.

Layer 3 – Executive Services Layer

This layer provides various services to the user. It interacts with input/output devices, object management, process management, and system security. This layer consists of various components. Various services are managed by each component. Various components of this layer are as follows:

■ Object Manager
■ Power Manager
■ Process Manager
■ I/O Manager
■ Virtual Memory Manager
■ Cache Manager
■ Security Reference Monitor
■ Plug and Play Monitor

Let us discuss each one of them in brief:

Object Manager: In a file system objects are files, directories, and folders. The object manager provides rules for creation, insertion, naming, and security of objects. It manages to remove the duplicate object resources.

Power Manager: It deals with power events like power-off, stand-by, and hibernate. Windows 2000 supports all of the latest standards in power management including the Advanced Power Management (APM) and Advanced Configuration and Power Interface (ACPI). Consequently, network devices can be powered off when not in use and dynamically reactivated when network access is required.

Process Manager: It manages the creation and deletion of processes. It provides a standard set of services for creating and using processes. It works along with the security model and the virtual memory manager to provide inter-process protection.

I/O Manager: It supports all file system drivers, hardware device drivers, and network drivers.

Virtual Memory Manager: The virtual memory manager maps virtual addresses in the process's address space to physical pages in the computer's memory. It hides the physical organization of memory from the processes. This ensures that processes do not access the memory of other processes.

Cache Manager: The cache manager changes the size of the cache dynamically when the amount of available RAM changes. When a process opens a file that is already present in the cache, the cache manager copies data from the cache to the process's memory, and vice versa, as read and write operations are performed.

Security Reference Monitor: It is responsible for enforcing the access validation. It controls the objects to access only those resources for which they have access rights.

Plug and Play Monitor: It indicates to the device drivers where to find various hardware devices such as modems, networks, sound cards, and so on. Its task is to match the physical devices with the software and to establish channels of communication between each physical device and its drivers.

Layer 4 – Environment Subsystems

Various types of applications can run in windows OS on the same graphical desktop. It runs applications for operating systems such as MS-DOS, OS/2, Windows, and POSIX. Windows supports a variety of applications through the use of Environment Subsystems, For example, through command prompt of Windows, you can get the CUI environment of DOS.

SUMMARY

In this chapter, you have learned about mobile operating systems. A mobile operating system is a system that operates a smartphone, tablet, PDA, and other mobile devices. This operating system is featured with touch screen, Bluetooth, Wi-Fi, GPS mobile navigation, video camera, voice recorder, music player, and so on. You have also learned about frequently used mobile operating systems like Android, iOS, and the Windows operating system. The Android operating system is based on the Linux Kernel from Google Inc., whereas iOS is from Apple Inc. Windows Phone usually operates with the Windows operating system and is from Microsoft, which provides various Microsoft services, such as OneDrive and Office, Xbox Music, Xbox Video, Xbox Live games, and Bing, and non-Microsoft services, such as Facebook and Google accounts to the users.

POINTS TO REMEMBER

- Android is a mobile operating system based on the Linux kernel and was developed by Google.
- The programs which control and communicate with the hardware are called drivers.
- Microsoft's own Bluetooth dongles have no external drivers and thus require at least Windows XP Service Pack 2.
- SQLite is the database engine used to display HTML content.
- Dalvik virtual machine is a type of JVM which is used to run applications in Android-based devices.
- The UIKit provides the basic tools required to implement graphical, event-driven apps in iOS.

REVIEW QUESTIONS

Q1. Write one feature of the Android operating system.

Q2. Describe the architecture of the Android operating system with the help of illustration.

Q3. Name the topmost layer in the Android operating system.

Q4. Discuss the features of the Android operating system.

Q5. Describe the architecture of the Android operating system.

Q6. Name the four abstraction layers of iOS.

Q7. Write the function of iAd.

Q8. Why are kernel-mode device drivers used?

Q9. Explain the concept of iOS.

Q10. Discuss the architecture of the Windows operating system.

CHAPTER 16

WINDOWS 10

INTRODUCTION AND DEVELOPMENT

Windows 10 is a personal computer operating system, released by Microsoft and officially unveiled in September 2014.

- **First version release:** October 2014
- **Consumer version release:** July 2015
- **Volume licensing release:** August 2015.

Windows 10 introduces a "universal" application architecture. It expands Metro-style apps that are designed to run across multiple Microsoft products. Its user interface was revised to handle transitions among a mouse-oriented interface and a touchscreen-optimized interface.

First release

It introduces a virtual desktop system, a window, Task View.

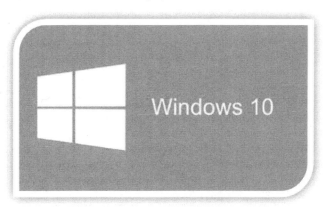

FIGURE 16.1. Windows 10

It receives an ongoing update to its features and functionality, and can receive non-critical updates at a slower pace.

The goal of Windows 10: To reduce fragmentation across the Windows platform, according to Terry Myerson, executive vice president of Microsoft's *Windows and Devices Group*.

In Windows 10 Microsoft downplays the user-interface mechanics that were introduced in Windows 8 in a non-touch environment, thus providing a desktop-oriented interface.

Development

- **2011:** At the Microsoft Worldwide Partner Conference, Andrew Lees stated that the company wants to have only a single software ecosystem for PCs, phones.
- **2013:** Microsoft was working on an update to Windows 8.
- **2014:** Updated version of Windows 8.1 which added an ability to run Windows Store apps inside desktop windows, was introduced by Terry Myerson.

Announcements

September 2014: Threshold was officially unveiled under the name *Windows 10*.

- According to Myerson, "Windows 10 would be Microsoft's 'most comprehensive platform ever,' which provides a single, unified platform for desktop computers, laptops, tablets, smartphones, and all-in-one devices."
- Windows 10 restores user interface mechanics from Windows 7, thus improving the experience for users on non-touch devices.

Release

June 2015: Microsoft announced that Windows 10 would be released in July 2015.

Preview releases

Public beta program or "Windows Insider Program" -- its preview is publicly available on October 1, 2014. Its users receive occasional updates and will continue to receive them after general availability (GA) in July 2015, in contrast with previous Windows beta programs.

Public release

- Windows 10 would be available (GA) on July 29, 2015, only before testing by vendors, because Microsoft was targeted to release this operating system in 2015 and the company was able to optimize its current and upcoming products in advance.

- Users can also in-place the upgrade through "Get Windows 10" application (GWX) and Windows Update that is identical in functionality to the Windows 8 online installer, and can be used to generate an ISO image.
- Changing among the architectures is not supported, and it requires a clean install.
- Windows 10 was available in 190 countries and 111 languages on its launch.
- Microsoft decided to become the partner with Qihoo and Tencent for promoting Windows 10.
- The price of Widows 10 is similar to the editions of Windows 8.
- Pro pack license of Windows allows easy upgrades of Windows 10 Home to Windows 10 Pro.

Editions and pricing

There are four main editions of Windows 10:

- Home versions for home users
- Pro version for small businesses and enthusiasts. It adds networking and security feature.
- Enterprise and Education version can be used for business environments, which is available only through volume licensing.

Free upgrade offer

First year of availability

- Upgrade licenses for Windows 10 are free for users who own a genuine license and installed the latest service pack.
- Enterprise customers who have an active Software Assurance (SA) contract with Microsoft can obtain Windows 10 Enterprise.
- According to November 2015 build, a free license of Windows 10 can be activated by using the product key of an existing Windows 7 or Windows 8.1 during installation.
- Insider Preview version of Windows 10 gets updated automatically to the released version.
- On June 2015, "Get Windows 10" application ("GWX") was activated on Windows devices running versions that are eligible to upgrade automatically, and also is compatible with, Windows 10.
- By using a system tray icon, users can easily access an application which advertises Windows 10.
- In July 2015, a pre-download process started which downloaded Windows 10 installation files to some computers that had reserved it.
- In September 2015, Microsoft triggers the automatic downloads of installation files of Windows 10.
- In October 2015, in Windows Update interface Windows 10 appear as an "Optional" update.

Licensing and Upgrades

- During upgrades, Windows 10 license status of the system's current installation is migrated.
- Digital entitlement is generated during the activation process.
- During the reinstallation process of Windows 10, if there have not been any significant hardware changes, then the online activation process will automatically recognize the system's digital entitlement in case no product key is entered during installations.

Updates and support

- Windows 10 delivery is described by Microsoft as a "service," because of its ongoing updates.
- On comparing with previous versions of Windows, Windows Update, selective installation of updates is not allowed.
- All the features updates and software driver updates are automatically downloaded and installed.
- With this, users can decide about their systems, regarding the reboot automatically for installing updates.
- By default, Windows Update uses a peer to peer system for distributing its updates.
- Initially, Microsoft stated Windows 10 would freely receive the updates for "supported lifetime of the device."

Upgraded builds

- Windows 10 upgraded builds will be released occasionally, and will contain new features and many other major improvements.
- The speed of receiving upgrades is dependent on the release channel on which it is used.
- Current Branch is the default branch for the users of Windows 10 Home and Pro, and it receives stable builds because they are publicly released by Microsoft.
- Windows Insider branches receive unstable build notices as they are released, at either a "Fast" or "Slow" pace.
- The Pro and Enterprise editions may or may not use the "Current Branch for Business" release channel.
- Windows 10 Enterprise may use the "Long-term support branch" (LTSB), which consist of periodic snapshots of Windows 10's Current Branch for Business branch, and receives only critical patches for over 10-year support lifecycle.

November 2015 upgrade

- November 2015 upgrade of Windows 10, also known as the "November Update," "version 1511", and "Threshold 2" (TH2).
- November 2015 upgrade distributed via Windows Update on November 12, 2015.

■ This November 2015 upgrade contains:
  • Improvements to the operating system
  • Improvements to the user interface
  • Many bundled services
  • Introduction of Skype-based universal messaging app
  • Windows Store for Business

Codename "Redstone."

■ After the release of Threshold 2, Windows Insider users get migrated to the "rs1_release" branch or "Redstone."
■ "Redstone" series is comprised of two major upgrades in 2016.
■ First release: build 11082, in December 2015.

System requirements to install Windows 10

Hardware requirements	Required	Recommended
Required Processor	1 GHz clock rate	x86-64 architecture
Memory	IA-32 edition: 1 GB x86-64 edition: 2 GB	4 GB
Graphics card	DirectX 9 WDDM	WDDM 1.3 or any higher driver
Display screen resolution	800×600 pixels	1024×768 pixels
Input device	Keyboard and mouse	Multi-touch display
Hard disk	IA-32 edition: 16 GB x86-64 edition: 20 GB	N/A

Features

■ Harmonizes user experience and functionality among the different classes of device.
■ Addresses the shortcomings in the user interface that were introduced in Windows 8.
■ The Windows Runtime app ecosystem was revised into the Universal Windows Platform (UWP) to run across multiple platforms and device classes.
■ Windows apps share code across platforms.
■ Windows Store also serves as a unified storefront for various apps.
■ Windows 10, a Goldilocks version of Microsoft's PC operating system and a kind of forward-looking touch screen vision of Windows 8.
■ Windows 10 is available as a free upgrade for the existing Windows 7 and 8 for the non-corporate users.
■ It is built from the ground up for pursuing the Microsoft's vision of unifying OS that spans on all devices without separating any platform.
■ Its development is a Microsoft's attempt for safe guarding Microsoft's crumbling software hegemony.

- It provides a single user experience that spans every piece of technology we touch.
- Windows 10 can be journey can be considered as a long, awkward road whose journey began with the release of Windows 8 in 2012, during the time when Microsoft was trying to convince a world of keyboard and mouse wielders that touch screens were the way to go.
- Windows 10 allows a smooth transition from "tablet" to "PC" mode.
- Live tiles that are present in the Windows 8 home screen still exist in Windows 10.

Review about Windows 10

- **Positive:** It bridges the gap among the PCs and tablets without excluding devices. This new OS combines the best features of old and new Windows into a package, thus resulting in the correction of all issues in Windows 8.
- **Bad:** Automatic, forced updates can lead to potential trouble later on.
- **Conclusion:** It gives a refined, improved vision for computing with an operating system that has a free upgrade.

User interface and desktop

- **Start menu:** It has a list of places, other options, and tiles that represent the application on the left and right side respectively.
- The menu can be resized.
- The virtual system was added.
- A new feature called Task View displays all open windows that allow users to switch between them.
- Windows Store apps can now be used in self-contained windows.

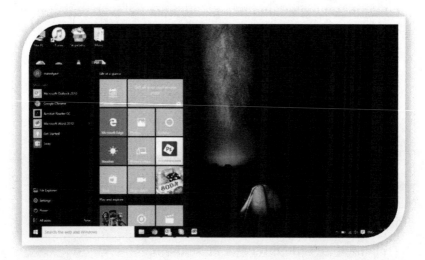

FIGURE 16.2. User interface and desktop

- Program windows can be snapped to quadrants by dragging them to the corner.
- Icons were changed to a minimalist design.
- *Charms* has been removed.
- Action Center is present; it continuously displays notifications and settings toggles.
- Offers two separate user interface modes, which are user interface optimized for mouse and keyboard, and a "Tablet mode" that are designed for touchscreens.

WINDOWS 10 SETTINGS MENU

The setting menu of Windows 10 includes basic setup, network settings, and privacy settings.

Features:

- Clean interface
- Bit
- Touch-friendly icons
- Toggles
- Simple descriptions

Start

- Charms → Settings → Change PC settings
- It can be located prominently on the Start menu and also it offers more functionality.
- In the Windows 10 menu, a user will find settings that never existed in the Control Panel, and these settings are *touchscreen* and *tablet*. Also, it includes icons that have been migrated from the Control Panel.

FIGURE 16.3. Setting menu

TABS IN THE NEW SETTINGS MENU:

■ **The System tab:** In Windows 10 a computer's basic settings can be changed in the new Settings menu. Its menu is robust, Control Panel-like version. Prompts such as system, devices, network & Internet and Update & security can be seen in the system tab.

FIGURE 16.4. System tab

■ **The Devices tab**

The Devices tab is a place where you can manage and fix any device issues. To find the Devices tab:

Start menu → Settings → Devices → Devices tab

FIGURE 16.5. Device tab

■ **The Network & Internet tab:** It combines the traditional Network and Sharing Centre from the Control Panel. Networks & Internet tab can be accessed:

Start → Settings → Network & Internet

Or

Network icon → Network settings.

FIGURE 16.6. Network & Internet tab

■ **The Personalization tab:** By using this, the user can change the desktop wallpaper, a color that appears throughout in Windows 10.

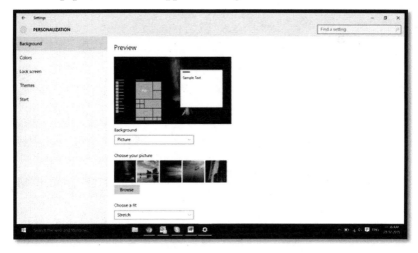

FIGURE 16.7. Personalization tab

■ **Accounts tab:** The accounts tab has five options: "Your email and accounts," "Sign-in options," "Work access," "Family & other users" and "Sync your settings."

FIGURE 16.8. Account tab

- **The Time & language tab:** By using this, the user can change the time and date, add languages, and adjust the speech settings.
  To go to Time and language tab:
  Go to Settings menu→click Time & language.

FIGURE 16.9. Time & language tab

- **The Ease of Access tab:** Under this tab, a high-contrast theme, voice narration, or closed captions can be found.
  For getting to command ease of access, go to:
  Charms bar → Settings → Change PC settings → Ease of Access
- **The Privacy tab:** Most of the device's basic privacy settings including location settings and diagnostic information can be found here.

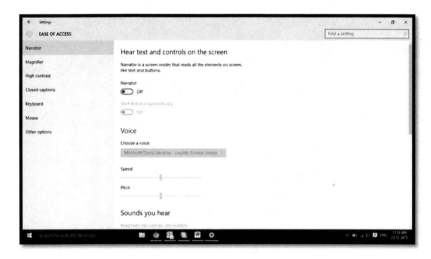

FIGURE 16.10. Ease of Access tab

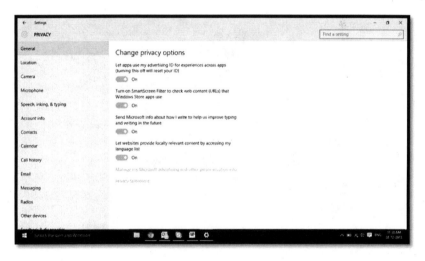

FIGURE 16.11. Privacy tab

■ **The Update & recovery tab:** Microsoft has placed Windows Update in the new Settings menu and removed it from the Control Panel.

THE START MENU

■ The start menu is back in Windows 10.
■ The start button is present in the lower left corner of the Windows desktop.
■ Start menu contains a column that contains shortcuts to apps.
■ On pressing the "All Apps" button, an alphabetical list of all apps that are installed on your PC appears.
■ There are colorful, animated, live tiles containing app shortcuts and informative widgets.

FIGURE 16.12. Update & recovery tab

FIGURE 16.13. Start menu

Other Start Menu Features

- It resembles the Windows 7 design as it uses only a portion of the screen and a Windows 7-style application listing in its first column.
- Second column displays Windows 8-style app tiles.
- A concept called, "universal Windows app," was also introduced, allowing Windows Store apps to be created for Windows 8.1 that can be ported to Windows Phone 8.1 and Xbox One.

WINDOWS 10 ENTERPRISE VERSION:

- In this version, an administrator can set up policies for the encryption of sensitive data.

- It also offers Device Guard, which allows administrators to enforce a high-security environment as it blocks the execution of software that is not digitally signed by a trusted vendor or Microsoft.
- It automatically compresses the system files for reducing the storage footprint of the operating system.

WEB BROWSERS

- *Microsoft Edge*, a new default web browser has been introduced in Windows 10(See Section 16.14).
- For compatibility purposes, Internet Explorer 11 is also maintained on Windows 10.
- It incorporates *Cortana*, which is a Microsoft's intelligent personal assistant; it replaced Windows' embedded search feature, and thus supports both text and voice input.
- It also offers a Wi-Fi Sense feature that originates from Windows Phone 8.1.

MULTIMEDIA AND GAMING

- Windows 10 provides heavier integration with the Xbox ecosystem that allows users to browse their game libraries.
- The Xbox Live SDK allows an application developer to incorporate Xbox Live functionality into their apps.
- Games such as *Candy Crush Saga* and *Microsoft Solitaire Collection* are bundled with Windows 10.

FIGURE 16.14. Multimedia and gaming

■ In Windows 10, Xbox Video and Xbox Music apps have been renamed to *Movies & TV* and *Groove Music*.

■ The function of Xbox Video and Xbox Music apps is identical; by using this the user can find any music and video files on a device, and it also serves as a means for either buying or renting the content from Microsoft's stores.

SHORTCUT KEYS

■ Windows 10 has numerous keyboard shortcuts.

■ It has touch gestures for each of the available features

Shortcut	Description
Windows key ⊞	Open and close the **Start** menu.
⊞+1, ⊞+2, etc.	Switch to the desktop and launch the nth application in the taskbar. For example, ⊞+1 launches whichever application is first in the list, numbered from left to right.
⊞+A	Open the action center.
⊞+B	Highlight the notification area.
⊞+C	Launch Cortana into listening mode.[1] Users can begin to speak to Cortana immediately.
⊞+D	Switch between **Show Desktop** (hides/shows any applications and other windows) and the previous state.
⊞+E	Switch to the desktop and launch File Explorer with the **Quick Access** tab displayed.
⊞+H	Open the **Share** ◙ charm.
⊞+I	Open the **Settings** ⚙ app.
⊞+K	Open the **Connect** pane to connect to wireless displays and audio devices.
⊞+L	Lock the device and go to the **Lock** screen.
⊞+M	Switch to the desktop and minimize all open windows.
⊞+O	Lock device orientation.
⊞+P	Open the **Project** pane to search and connect to external displays and projectors.
⊞+R	Display the **Run** dialog box.
⊞+S	Launch Cortana.[2] Users can begin to type a query immediately.
⊞+T	Cycle through the apps on the taskbar.
⊞+U	Launch the Ease of Access Center.
⊞+V	Cycle through notifications.
⊞+X	Open the advanced menu in the lower-left corner of the screen.

FIGURE 16.15. Shortcut Keys in Windows 10

⊞+Z	Open the app-specific command bar.
⊞+ENTER	Launch Narrator.
⊞+SPACEBAR	Switch input language and keyboard layout.
⊞+TAB	Open Task view.
⊞+,	Peek at the desktop.
⊞+Plus Sign	Zoom in.
⊞+Minus Sign	Zoom out.
⊞+ESCAPE	Close Magnifier.
⊞+LEFT ARROW	Dock the active window to the left half of the monitor.
⊞+RIGHT ARROW	Dock the active window to the right half of the monitor.
⊞+UP ARROW	Maximize the active window vertically and horizontally.
⊞+DOWN ARROW	Restore or minimize the active window.
⊞+SHIFT+UP ARROW	Maximize the active window vertically, maintaining the current width.
⊞+SHIFT+ DOWN ARROW	Restore or minimize the active window vertically, maintaining the current width.
⊞+SHIFT+LEFT ARROW	With multiple monitors, move the active window to the monitor on the left.
⊞+SHIFT+RIGHT ARROW	With multiple monitors, move the active window to the monitor on the right.
⊞+HOME	Minimize all nonactive windows; restore on second keystroke.
⊞+PRNT SCRN	Take a picture of the screen and place it in the **Computer>Pictures>Screenshots** folder.
⊞+CTRL+LEFT/ RIGHT arrow	Switch to the next or previous virtual desktop.
⊞+CTRL+D	Create a new virtual desktop.
⊞+CTRL+F4	Close the current virtual desktop.
⊞+?	Launch the Windows Feedback App.

[1] If Cortana is unavailable or disabled, this shortcut has no function.
[2] Cortana is only available in certain countries/regions, and some Cortana features might not be available everywhere.
[3] If Cortana is unavailable or disabled, this command opens Search.

DIRECTX 12

- Windows 10 includes DirectX 12 which provides "console-level efficiency" with "closer to the metal" features and thus has a flexible access to hardware resources, and also reduces CPU and the graphics driver overhead.
- Its performance improvements can be achieved through low-level programming, which allow developers to use the resources in an efficient way.
- DirectX 12 also supports vendor-agnostic multi-GPU setups.
- Windows 10 also includes WDDM 2.0, which provides a new virtual memory management and allocation system for reducing the workload on the kernel-mode driver.

FEATURES REMOVED IN WINDOWS 10

- OneDrive, which was the built-in sync client, was introduced in Windows 8.1; it no longer supports offline placeholders for online-only files in Windows 10.
- Viewing of offline files is expected to be added in a new Windows app.
- Users cannot synchronize with the Start menu unless associated with a Microsoft account.
- Automatic installation of a Windows Store was also removed.
- Web browsers cannot set user's default without further intervention.
- Default web browsers can be changed manually from the Settings' "Default Apps" page.
- The MSN applications such as *Food & Drink, Health & Fitness,* and *Travel* have been discontinued.
- Changes that are made within an individual update are now withheld by Microsoft.

VIRTUAL DESKTOPS

- Virtual desktops can be obtained by:
  By clicking the Task View button present on the taskbar → bird's-eye view of all apps get open → Drag any one apps on the "new desktop" button → it will move independently on a workspace.
- Virtual desktops help you spread your apps across several workspaces.
  The user can move their apps across virtual desktops by simply dragging them, or by right-clicking

WINDOWS SNAP

- A snap feature introduced in Windows 7 has been updated in Windows 10.
- It shows you the thumbnails of all other apps that are currently open.
- The user can also snap an app into a corner of the screen.

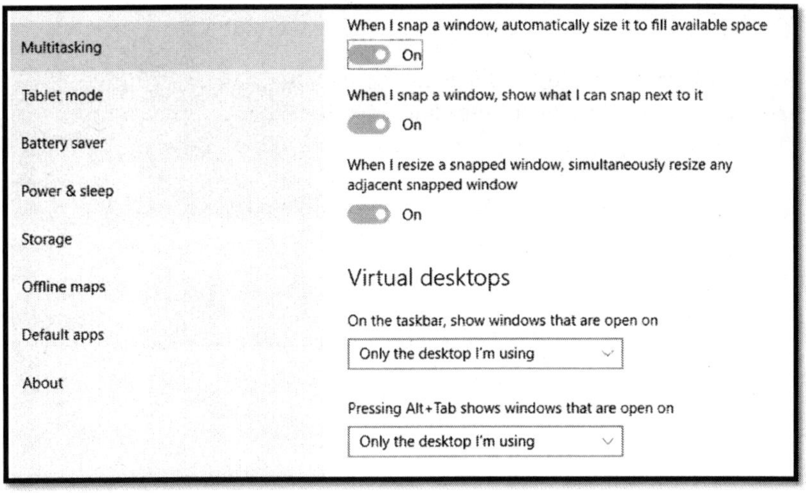

FIGURE 16.16. Virtual desktop

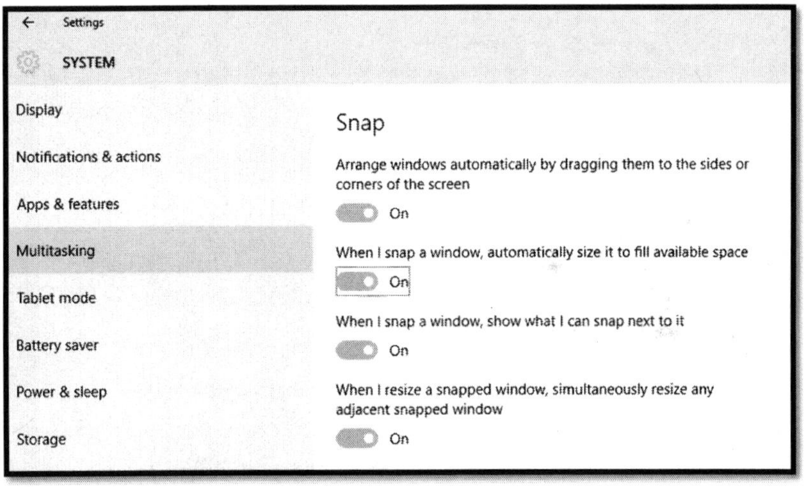

FIGURE 16.17. Window snap

Action Center

■ Action Center of Windows 10 replaces the "Charms" which was introduced in Windows 8.

■ It is a nod to mobile operating systems.

SYSTEM AND SECURITY

Windows 10 employs a multi-factor authentication technology, as its operating system includes improved support for biometric authentication.

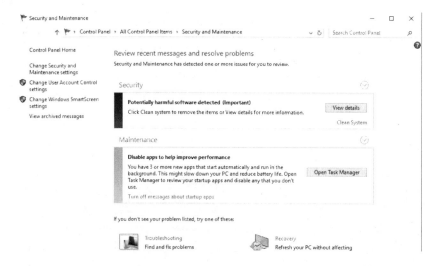

FIGURE 16.18. System and security

Windows Hello and Windows Passport

For security purposes, Windows 10 uses the devices' cameras or a finger-print scanner. Once a user has authenticated itself with *Windows Hello*, then *Windows Passport* gives the user access to a third-party sites and products. Now there is no need for the user to log in again and again.

MICROSOFT EDGE: A NEW BROWSER

- A brand-new browser has been added by Microsoft into Windows 10, which is called *Microsoft Edge*.
- Edge is a fast, modern browser
- *Cortana* is integrated into the browser. It offers detailed information on various things like weather forecasts, news, etc.
- Cortana can pull up a little sidebar full of useful information, like reviews or directions when the user is searching for something.
- Another feature that Edge offers is *web note* that allows a user to scribble on web pages.
- The "reading view" option can also be used by a user to scroll down a website.
- Edge has been provided with tighter security.

PHOTO APPS

- *The photo app* is used to make quick and a non-destructive edit to a user's pictures.
- It scans the devices and OneDrive account for photos, and then automatically arranges pictures into albums.
- This app is used to keep the track of your pictures, and also offers the user some basic editing tools.

FIGURE 16.19. Photo App

- Photo app will automatically enhance all the photos as it will automatically fix red eye.
- Photo app can also work on RAW files.

WINDOWS, EVERYWHERE

- This version of Windows is different from any other Windows version.
- Microsoft will be promoting Windows 10 features related to the mobile and universal apps.
- Universal apps of Microsoft share an identical code base that can be the Excel client present on your desktop.
- Universal apps will also lead to challenges, as a developer has to create a rich, robust app that runs on a mobile device, versus developing apps have the usage of the power of a PC.

GETTING READY FOR THE NEXT UPDATE

- The Windows update process is one of its most contentious elements.
- The system updates automatically.
- It becomes a prime target for malware.
- One can delay the updates, but can't avoid them.

DRAWBACKS OF WINDOWS 10

Windows 10 has been criticized for the following reasons:

- Limiting the number of users to control its operation.
- It installs all updates automatically, except its *Pro* edition.
- Its settings require the transmission of user data to Microsoft or its partners.

REVIEW QUESTIONS

Q1. List two key features of the Windows 10 operating system.
Q2. Describe the architecture of the Windows 10 operating system.
Q3. What is the name of the new Microsoft browser found in Windows 10?
Q4. What is Cortana?
Q5. Where is the START menu found in Windows 10?

INDEX